Clay Tech: Layered Silicates Boost Polymers

Mahima

Contents

Contents

Contents

Chapter 1

Abstract

Polymer-nanocomposites based on sustainable fillers are integral to the development of modern high-performance materials. Consequently, the successful preparation of Montmorillonite (MMT)–polymer-nanocomposites with tailorable mechanical properties *via* surface grafting of RAFT polymers was established. Strategies to precisely tune the properties of filler–matrix and matrix-free composites were developed and characterized *via* tensile testing. The possible application of these approaches for the synthesis composites with enhanced gas barrier properties was explored.

The naturally occurring layered silicate MMT was modified with an ionic monomer for radical polymerization *via* an ion exchange procedure and subsequently employed in a *grafting through* RAFT-polymerization of methyl acrylate. The successful surface modification with monomer and polymer was demonstrated with a variety of methods. Thermogravimetric analysis (TGA) and elemental analysis (EA) both showed a significant increase of organic content after modification. attenuated total reflectance Fourier-transform infrared (ATR-FTIR) spectroscopy was used to demonstrate the presence of all components (MMT, surface grafted monomer, and polymer) in the resulting composites. scanning near-field optical microscopy (s-SNOM) allowed for the proof that a significant amount of polymer can be immobilized at the surface while any unbound polymer can be separated from the sample. Hence, the production of hybrid nanosheets only consisting of MMT and surface-tethered polymer was achieved. Small angle X-ray scattering (SAXS) was employed to show the successful intercalation of monomer and polymer in between the MMT layers. This results in exfoliation of the nanoclay and enables access to a polymer modified filler with a high aspect ratio. This was further supported by transmission electron microscopy (TEM) which provided visual evidence of the change in particle morphology induced by the polymer modification.

The first type of nanocomposites that was studied consisted of a poly(methyl acrylate) (PMA) matrix filled with nanosheets to reinforce the composite material. First of all, composites filled with nanosheets with either no modification, monomer modification, or polymer modification were compared. It could be demonstrated that the strongest reinforcement is indeed observed when polymer coated nanosheets are used. Subsequently, the effect of the nanosheet content on the mechanical properties was studied. A nearly linear dependence of *Young's* modulus and tensile strength on the filler content has been observed. The strain at break is correlated anti-proportionally to the filler content while the toughness exhibits no dependency on the MMT content. In comparison with dynamic mechanical analysis (DMA) this effect could be attributed to a reduced chain mobility of the polymer as consequence of the formation of a glassy layer around the particles. Additionally, the effect of the grafted polymers chain length on the mechanical performance was explored. This strategy also resulted in an enhanced modulus and tensile strength that linearly depend on the grafted chain length. It was characteristic for these composites that the plastic deformation stayed unmodified resulting in a significantly enhanced overall toughness. The observed effects were attributed to enhanced entanglement of the surface bound chains with the matrix chains resulting in improved stress propagation. Microscopy studies resulted in further insight into the behaviour of the nanosheets under external stress, revealing that individual particles break in tensile direction. It could also be shown that this has no negative impact on the materials performance.

The second type of studied nanocomposites are matrix-free composites consisting only of nanosheets and surface-bound polymer without any free polymer chains. Using this technique, much higher nanosheet contents are accessible. As evident from this study, the formation of coherent and mechanically loadable composites requires a minimal grafted chain length in order to allow chains to entangle with each other. Variation of the grafted polymer chain length could be used as a tool to adjust the MMT content and to tune the composites properties. It was observed that *Young's* modulus and tensile strength linearly increase with decreasing grafted chain length while the strain-at-break and toughness decrease simultaneously. It could further be demonstrated that the addition of only a small proportion of nanosheets decorated with polymer of a high molar mass is sufficient to allow film formation of particles modified with polymer of a low molar mass. This strategy enables the synthesis of films with even higher MMT content. Cross-linking of the polymer *via* addition of star polymer or the introduction of hydrogen bonding moieties in the polymer allowed for further reinforcement of the composite.

To explore possible applications of the described techniques it was demonstrated that poly ethylene foils coated with PMA decorated nanosheets exhibit enhanced mechanical properties as well as enhanced gas barrier properties.

Chapter 2

Introduction and Motivation

Polymers are reinforced with a variety of particles in order to tailor their mechanical,[1–5] thermal,[2,3,6] electrical,[1,7] and optical[8,9] properties among others. Two materials with opposing properties, for example hydrophilic-hydrophobic,[10,11] organic-inorganic,[10,12] or ductile-stiff,[13,14] are combined resulting in a superior material that exhibits traits of both materials. This approach shows extraordinary versatility, as all known polymers can be combined with a great diversity of particles ranging from natural fillers, such as layered silicates[1,15,16] or wood particles,[17] over synthetic particles, like silica,[18,19] carbon black[20] or carbon nanotubes,[21] to metal particles like gold[7,22,23] and silver[7,23] or metal oxides, like magnetite.[4] In doing so, an exceptionally broad range of materials can be created.

Filler materials can be characterized by their spacial dimensionality ranging from zero dimensional fillers, like for example spherical silica nanoparticles,[19] over one dimensional fillers, like carbon nanotubes,[24] to two dimensional fillers like graphene[24] or layered silicates.[1,25,26] While all of these fillers have been commercially applied, especially layered silicates have been the focus of extensive research over the past 30 years due to their high aspect ratio, which makes them promising for mechanical reinforcement of polymer matrices.[27,28] In the beginning of the 1990s, the Toyota group presented the first clay/nylon-6-6 composite.[29–31] Since then, tremendous research efforts were focused on various polymer–layered-silicate–

nanocomposites (PLSNs).

One of the most commonly used layered-silicates for polymer modification is M MT.[32] Even tough it is easily available at a low cost, the incorporation of the hydrophilic silicate into hydrophobic polymer matrices remains challenging.[33] Sophisticated surface modification is required to enhance the miscibility of the silicate and the polymer matrix.[34] Many approaches to achieve this focus on ionic surface modification which allows for cleaving of the surface modification if desired.[11,35,36] Combined with the fact that MMT, unlike for example silica[37] or carbon black,[38] can be mined easily and little to no resources are needed to make it available for commercial applications, this makes these materials also interesting in the context of "green chemistry" as they are potentially recyclable and reusable.[39,40]

When taking a look at the historical development of materials, it becomes clear that new materials were mostly created by trial and error or by chance. First craftsmen and later scientists tried to combine materials in order to create new materials and only afterwards analyzed them and evaluated their use.[41] In modern times, with modern technology aiming higher and higher, the situation is often the other way around: For a designed application a suitable material with predetermined characteristics needs to be developed. In order to achieve this, the effect a certain filler has onto a polymeric carrier matrix needs to be understood and characterized. The scope of this book is the synthesis poly(acrylate)–MMT-nanocomposites and the development of strategies to precisely tailor their mechanical properties.

The first part of this book focuses on the surface modification of MMT with PMA. In order to fully characterize the filler material, a wide rage of methods will be applied. This provides the reader with a variety of possibilities to chose from for future projects. The application of size exclusion chromatography (SEC), Thermogravimetric analysis (TGA), IR-spectroscopy, scanning near-field optical microscopy (s-SNOM), elemental analysis (EA), transmission electron microscopy (TEM), and Small angle X-ray scattering (SAXS) for the thorough analysis of polymer grafted MMT will be presented. In the second part of the book, the performance of polymer grafted MMT as nanofiller in a polymeric matrix is demonstrated through a systematic study of the mechanical properties of the composite. This is followed by the development of a biomimetic composite material of MMT and both PMA and poly(butyl acrylate) (PBA) which is based on the microscopic structure of nacre and is designed without the use of any matrix polymer. The strategy how to successfully produce stable thin films from this material will be developed within this book and is complemented by different approaches to finely tune the mechanical properties of the product. Lastly, a possible application of the presented nanocomposites is explored. poly(ethylene) (PE)-foils are coated with hybrid polymer–MMT-nanosheets and are analyzed regarding their mechanical and gas barrier properties.

Chapter 3

Theoretical Background

Contents

Within this chapter, the background knowledge to this book is presented. Montmorillonite, a naturally occurring layered silicate which is used throughout this book as a filler material for polymers, is introduced and described. Furthermore, some non well-known methods that have been used as key techniques within this book are presented.

A key polymerization technique used throughout this book is the radical addition–fragmentation chain transfer (RAFT) polymerization. Tensile testing and DMA that are used to evaluate the change in mechanical properties and SAXS that provides inside to the change in particle properties during functionalization are shortly explained.

Furthermore, a general overview on polymer-nanocomposites and different contributions to the mechanical performance of these materials is given. All techniques and concepts are explained to the point necessary to understand the experiments and results presented and discussed in this book. For further details, the reader is referred to appropriate literature.

3.1. Methods used throughout this book

3.1.1. RAFT polymerization on surfaces

All polymers used throughout this book were synthesized *via* RAFT Polymerization. In a typical RAFT procedure the mechanism of a conventional radical polymerization is superposed with a RAFT equilibrium induced by an added chain transfer agent.[42,43] Scheme 3.1 shows the general structure of a RAFT agent.[42] The Z-group is a stabilizing group that provides stability to the intermediate radical (Scheme 3.2) while the R-group is the leaving group. The R-group needs to be able to fragment quickly and re-initiate a polymerization. R- and Z-group can be chosen to suit the used monomer well.[43] Upon initiator decomposition and a few steps of monomer addition to the initiator radicals the growing radicals can add to a RAFT agent resulting in the depicted RAFT equilibrium. The intermediate RAFT agent radical can fragment to both sides of the reaction equation which leads to a fast and efficient exchange of all growing polymer chains with the present RAFT molecules. This results in a narrow molecular weight distribution and allows full control over the molecular weight as the conversion is linear with time.[42,43] Additionally, this opens up the possibility of the formation of different polymeric architectures, as for example block copolymers, because the RAFT groups stay covalently linked to the polymer and can be reactivated in a second polymerization step with for example a different monomer. Further-

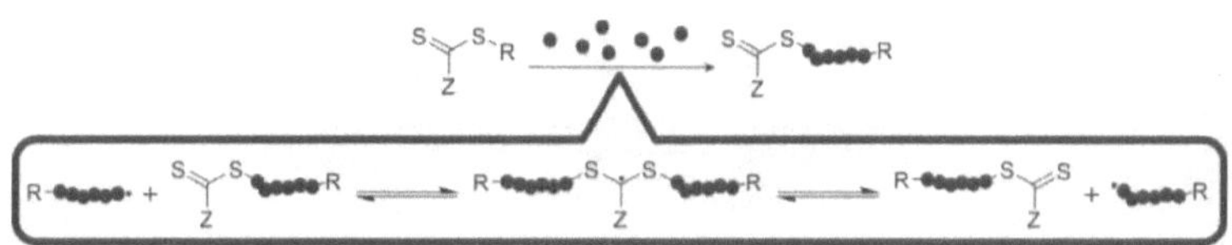

Scheme 3.1: Schematic structure of a RAFT agent.[42]

Scheme 3.2: General mechanism of a RAFT polymerization.[42] Blue balls represent monomer units.

more, variation in the geometry of the RAFT agent gives access to different polymer architectures as for example star polymers.[43]

There are different strategies to use the RAFT mechanism to synthesize polymers on surfaces. They can be divided into three classes (Scheme 3.3): *grafting to*, *grafting from*, and *grafting through*.[44] In a *grafting to* approach the polymers are synthesized and then in a second step grafted to the surface. Surface and polymer have to exhibit suitable functionalities for binding the polymer to the surface. This strategy has the clear advantage that the polymer can be fully characterized beforehand.[43,44] Regarding the MMT this is not a suitable approach as the produced polymer chains are sterically demanding and are therefore not able to diffuse easily in between the MMT layers. *Grafting from* is a very elegant approach where either an initiator or the RAFT agent is anchored to the surface. As the respective compound is part of the later polymer chain, these chains are directly bound to the surface. When a RAFT agent is bound to the surface, it can be anchored either *via* its R-group[45–47] or *via* its Z-group.[47–50] If the R-group is anchored, the growing chain is bound to the surface while the fragmented RAFT group is in solution. In consequence, the propagating radicals are found on the outer sphere

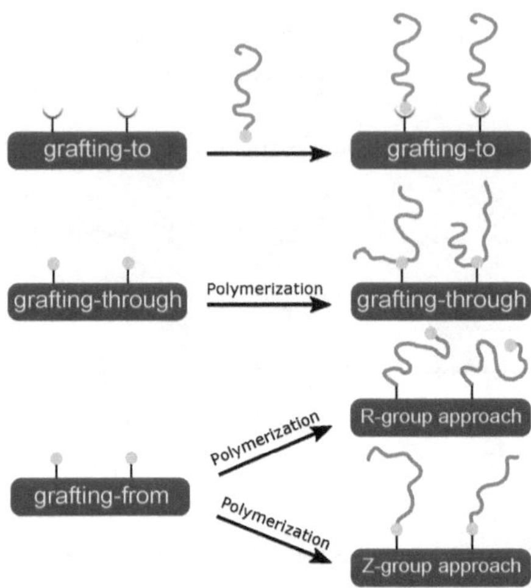

Scheme 3.3: Schematic representation of *grafting to, grafting through,* and *grafting from* approaches. Yellow sheres represent the for the mechanism characteristic functionalities: functionalities that can react with a surface located functionality (*grafting to*), surface anchored monomer (*grafting through*), or surface bound initiator or control agent (*grafting from*).[44]

of the growing polymer layer. They are therefore easily accessible for monomer which allows for high grafting densities.[45] If the Z-group is bound to the surface, the RAFT agent is anchored to the surface and the propagation step occurs in solution.[49,50] In order to add to a RAFT, agent the active radical has to diffuse to the RAFT agent. With increasing degree of polymerization this diffusion is hindered both by the increasing size of the radical chain and by a shielding effect originating from other polymer chains bound to RAFT agents in close proximity. This effect allows only for lower grafting densities.[43–45] In the case of MMT, however, to anchor a RAFT agent to the surface a large excess of RAFT agent would be needed to modify the

MMT with RAFT agent which would be a very expensive and uneconomic approach where most of the RAFT agent would be wasted.[51] Therefore, the approach that was used throughout this book is a *grafting through* approach.[39] Here, a monomer is bound to the surface. Compared to the anchoring of a RAFT agent this has the advantage that suitable monomers are inexpensive as well as easy and fast to synthesize. The RAFT mechanism is carried out in solution in the presence of the monomer modified particles. At some point the growing radicals will statistically include a surface bound monomer unit into the growing chain. At this point, the chain will be bound to the surface.[52–54] If the ratio of RAFT agent to surface bound monomer is kept larger than one additionally to surface bound polymer also free polymer will form. This can be either separated from the polymer modified particles *via* centrifugation or it can directly be considered matrix polymer.

3.1.2. Tensile testing[41,55,56]

Tensile testing is a well-established mechanical testing method with the purpose to determine certain materials properties that allow for predictions on how the material behaves under external stresses. Knowledge on a materials characteristics is crucial to evaluate a materials possible application.

In a uni-axial tensile testing a typically dog-bone shaped sample is elongated at a constant strain rate while the necessary force is detected. The relative elongation of the sample is called the strain (ε) and is given as the absolute elongation of the sample l relative to its initial length l_0:

$$\varepsilon = \frac{l - l_0}{l_0}. \tag{3.1}$$

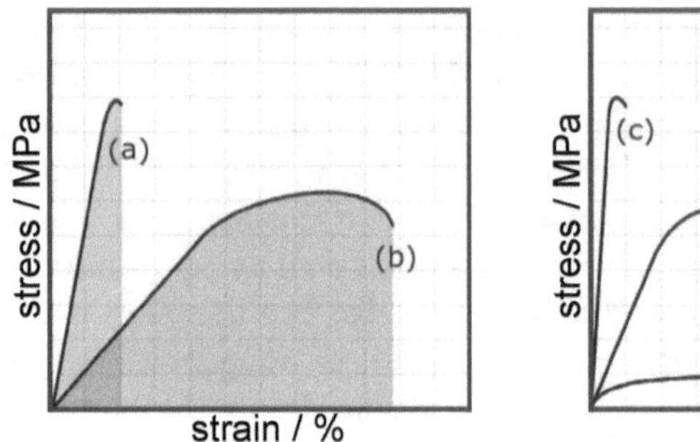
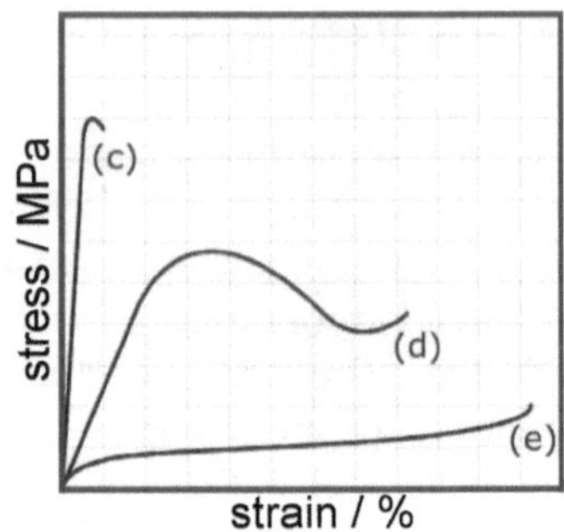

Figure 3.1.: Schematic stress–strain-curves. The stress is depicted in dependency of the strain. The left hand side shows stress–strain-curves of a brittle (a) and a ductile (b) material. Their respective toughnesses are highlighted as the area under the curves. On the right hand side characteristic stress–strain-curves for three different classes of polymers are shown: duroplastic (c), thermoplastic (d), and elastomeric (e).

Division of the measured force (F) by the initial cross section (A_0) results in the stress (σ):

$$\sigma = \frac{F}{A_0},\tag{3.2}$$

that can then be depicted in dependence of the strain in a so-called stress–strain-curve. The cross section of the sample decreases over the course of the experiment. The stress σ is nevertheless calculated *via* division by the initial cross section. The resulting values do therefore not reflect the true stress but rather an engineering stress which is always smaller than the true stress. Exemplary stress–strain-curves are shown in Figure 3.1 (a) and (b).

In the region of low strains all deformation is reversible. Here, the relation between stress and strain is nearly linear and is described *via Hooke's Law*:

$$\sigma = E \cdot \varepsilon, \tag{3.3}$$

with the proportionally constant being the *Young's* modulus (E). The highest stress that is reached is called the tensile strength (σ_y). The linear region progresses into a non-linear region where the testing specimen is deformed plastically and eventually breaks giving access to two further characteristic values: the stress at break (σ_B) and the strain at break (ε_b). The area under the stress–strain-curve is a measure for the energy absorbed by the sample until the break and is called the toughness (U_T) (compare Figure 3.1 (a) and (b), colored area).

Polymers are usually divided into three main classes: Duroplasts, thermoplasts, and elastomers. Figure 3.1(c)-(e) show exemplary stress–strain-curves for these three classes of polymers. Duroplasts show a high *Young's* modulus and a high tensile strength. They also show nearly no plastic deformation. They are highly cross-linked polymers, non-meltable, and usually brittle. Thermoplasts are linear non-cross-linked polymers. This enables them to flow when an external force is applied. Subsequent to the the linear region in the stress–strain-curve plastic deformation is observable. After elastic deformation the polymer chains flow past each other and are oriented during this process. Consequently, nearly no increase in stress is observed while the strain increases dramatically. When all chains are oriented, the stress rises again until the sample breaks. Elastomers, finally, are loosely cross-linked polymers. They also show elastic and plastic deformation, typically up to very high strains, but regain their shape after the external force is taken away.

3.1.3. Dynamic mechanical analysis[57]

DMA is a non-destructive testing method which is used to study viscoelastic properties of materials. A sinusoidal strain is applied to a sample and the needed force is measured. Measurements are typically recorded isothermal-frequency dependent or isofrequent-temperature dependent.

As all non-ideal materials, polymers have partly linear elastic and partly viscous properties. The complex modulus (E^*) of such a material can be written as

$$E^* = E' + iE'',\qquad(3.4)$$

with the storage modulus (E') and loss modulus (E''). The storage modulus describes the part of the energy that is brought into the system as mechanical energy that is stored and is regained when the applied stress returns to zero. The loss modulus describes the part of the energy that is dissipated as heat resulting from internal friction.

During a dynamic-mechanical experiment a sinusoidal strain

$$\varepsilon(t) = \varepsilon_0 \cdot \sin(\omega t),\qquad(3.5)$$

is imposed on the specimen. It depends on the time dependent strain ($\varepsilon(t)$), the amplitude of the oscillation (ε_0), and the applied frequency (ω). The detector analyzes the needed stress for a set strain resulting in a periodical stress

$$\sigma(t) = \sigma_0 \cdot \sin(\omega t + \delta),\qquad(3.6)$$

with the time dependent stress ($\sigma(t)$), the stress amplitude (σ_0), and the phase shift between deformation and stress (δ).
For purely elastic materials δ results to zero, while purely viscous materials show a phase shift of $\delta = 90°$. Polymers have both linear and viscous properties and do therefore show a phase shift $0° < \delta <$

90°. Combination of Equation 3.5, Equation 3.6, and Equation 3.3 gives an expression for the complex modulus:

$$E^* = \frac{\sigma_0}{\varepsilon_0} \cos \delta + i \frac{\sigma_0}{\varepsilon_0} \sin \delta. \tag{3.7}$$

Comparison with Equation 3.4 gives expressions for the storage E' and loss modulus E'':

$$E' = \frac{\sigma_0}{\varepsilon_0} \cos \delta \quad \text{and} \quad E'' = \frac{\sigma_0}{\varepsilon_0} \sin \delta. \tag{3.8}$$

The quotient of loss and storage modulus is called the loss factor $\tan(\delta)$ and written as

$$\tan(\delta) = \frac{E''}{E'}. \tag{3.9}$$

Figure 3.2 shows an exemplary temperature-dependent DMA curve exhibiting both an alpha- or glass-transition and a beta transition. At low temperatures the amorphous polymer is in a glassy frozen state. The polymer behaves like a solid and its response to an external force is mainly elastic. This results in a high storage modulus and a lower loss modulus. In consequence, the loss factor is low. When the temperature rises, the free volume increases and side chain mobility is possible. Additionally, short parts of the backbone become mobile and can move in a "crankshaft"-like motion. Both can be origins of secondary transitions. As the temperature rises the free volume of the polymer increases further. At the glass transition long range motion of entire polymer chains is possible. This is indicated by a significant drop in storage modulus of up to three orders of magnitude. The energy that is brought into the system as heat is used for chain movement. As a result of internal friction, energy is lost and the loss modulus exhibits a peak. In consequence of both,

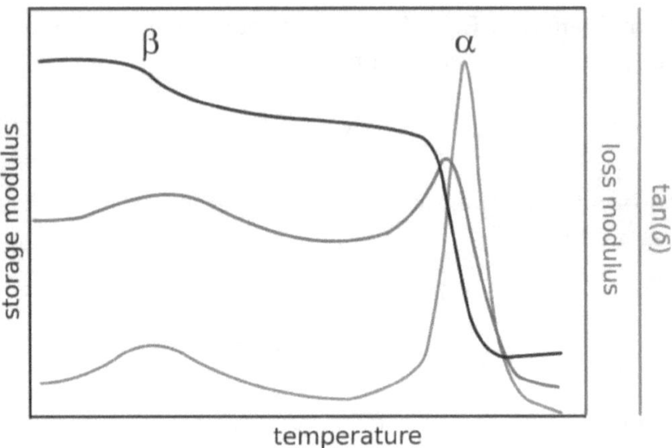

Figure 3.2.: Schematic DMA curve. Shown are the storage modulus E', the loss modulus E'', and the loss factor $\tan(\delta)$ in dependency of the Temperature T. α and β represent the position of an alpha- and a beta-transition.

the loss factor exhibits a peak as well. In literature, transitions are usually labelled with Greek letters going towards lower temperatures from the glass or alpha transition.

3.1.4. Characterization of MMT-nanocomposites *via* small angle X-ray scattering

X-Ray scattering techniques allow the investigation of a sample's properties on the nanometer scale. They are applied for various purposes such as the detection of crystal structures, chemical composition, or different physical properties, as residual stress or texture.[58] SAXS focuses on the scattering with a diffraction angle (Θ) of usually less than 10°. This allows the analysis of large scale structures from nanometers to micrometers. SAXS is widely used to characterize polymers, proteins, colloids, structural defects as well as structures in amorphous and crystalline materials.[59–61]

Generally, *Bragg*'s law,

$$n\lambda = 2d \sin \theta, \tag{3.10}$$

describes the relation between Θ and the distance (d), the scattering order (n), and the wave length (λ) of the used radiation. The distance could for instance be the distance between two lattice planes (in WAXS) or between two layers of MMT (in SAXS). This means that for the investigation of MMT scattering techniques have two major purposes: wise angle X-ray scattering (WAXS), on the one hand, gives the possibility to determine the crystal structure of the MMT-Layers. SAXS, on the other hand, gives insights to the inter layer spacing, meaning the distance of the MMT layers. The naturally occurring MMT is surface decorated with typically sodium cations. As this distance is very uniform and periodically, a scattering peak is observed.[62] The sodium cations can be exchanged with other cations. If these are more sterically demanding then the sodium cations, the distance between the layers will increase. This increase can be monitored by SAXS.[9,39]

3.2. Tuning the properties of polymer-nanocomposites

The mechanical properties of unfilled polymers are mostly defined by their cross-link density, the flexibility of their backbone, and the temperature. To tune the mechanical properties of polymers, the incorporation of particles into the polymeric matrix is a well-known strategy to achieve outstanding results.[17,63–65] This approach shows an extraordinary versatility as all known polymers can be combined with an extreme diversity of particles ranging from natural fillers, such as layered silicates[1,15,16] or wood particles,[17] over synthetic inorganic particles, as silica,[18,19] carbon black or carbon nanotubes,[21] to metal particles, like gold,[7,22,23] silver[7,23] or metal oxides, like magnetite.[4] In doing so, an exceptional broadness of new materials can be created.

Filler materials can not only be classified by their origin and chemical properties but also by dimensionality. Zero dimensional fillers, like various spherical nanoparticles, one dimensional fillers, e.g. carbon nanotubes, and two dimensional fillers, as graphene nanosheets or layered silicates have been successfully used to tune the properties of polymers.[66] Amongst them, two dimensional sheet-like fillers are known to have the highest influence at comparable filler content on the mechanical properties of composites as they exhibit the highest aspect ratio.[27] *Yudin et al.* compared the influence of MMT nanosheets, silicate nanotubes, and spherical zirconium dioxide nanoparticles on the mechanical properties of a polyimide matrix.[27] They showed that at constant filler loading the strongest enhancement of the *Young's* modulus and the tensile strength was observed for composites filled with MMT. PLSNs have since been shown to exhibit higher moduli,[27,67] strength,[68,69] and thermal resistance,[15,70] as well as increased barrier properties,[26,71] and flame retardancy[14] compared to the respective unfilled polymers.

The effect of particles on the mechanical properties is a field of intense research and development and especially for layered silicates the reinforcing mechanism is not fully understood yet.[72] Most research so far has focused on the reinforcement mechanism in elastomers as they are most relevant for technical application for example in the tire industry. Even less in known about the reinforcement of thermoplastic polymers. However, in a first approach the developed theories should also give an idea for the mechanism in non cross-linked polymer-nanocomposites.

A first step to understand how fillers influence the mechanical

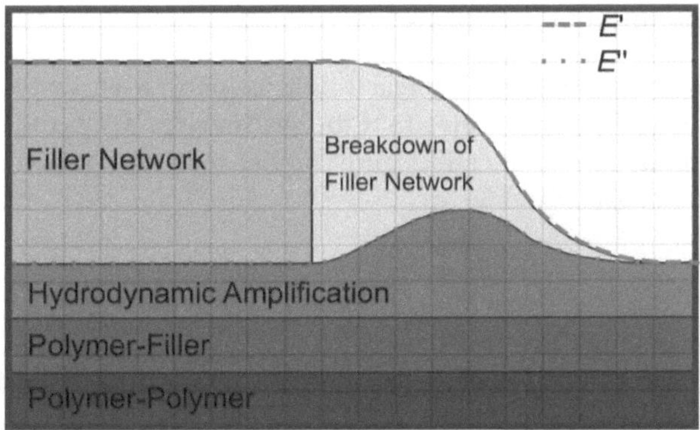

Figure 3.3.: Different interactions in Polymer–Filler-Nanocomposites and their contribution to storage and loss modulus with increasing strain. Adapted from *Raghunath et al.*[76]

properties of polymers is to evaluate different possible interactions that can occur in such a material. Here, four main contributions have to be taken into account:[73–75]

- hydrodynamic interactions

- filler–filler interactions

- polymer–polymer interactions

- polymer–filler interactions.

Figure 3.3 sums up the different reinforcing contributions to both storage and loss modulus and their dependency on an external strain. It is empirically known how storage and loss modulus behave with increasing strain. Different approaches have been made to explain all the contributions and effects. They each will be discussed in more detail over the further course of this chapter.

Hydrodynamic interactions

The first theory on mechanical reinforcement was postulated by *Einstein* in 1906[57] and states that the viscosity of a fluid filled with spherical particles of a significantly higher viscosity than the one of the fluid behaves as

$$\eta_\Phi = \eta \cdot (1 + 2.5 \cdot \Phi), \tag{3.11}$$

with η_Φ being the overall viscosity of the system, η the viscosity of the fluid and Φ the filler content. This concept is called hydrodynamic reinforcement and is based on the assumption of no interaction between fillers and fluid.[57,77] Using the correlation of modulus and viscosity results in:

$$G_\Phi = G \cdot (1 + 2.5 \cdot \Phi). \tag{3.12}$$

This correlation is also known as Einstein–Smallwood equation.[64] The decrease in storage modulus and the observed peak for the loss modulus with increasing strain is in this theory only attributed to the matrix.[57] While this theory does not take any interactions between the components into account, it does nevertheless describe one aspect of mechanical reinforcement.

Filler–filler interactions

Studies on the dependency of the modulus of a composite material on the filler content have shown that, contrary to the intuitive thought, a perfect dispersion of particles does not results in the best reinforcement but that the formation of a filler network is favorable for the mechanical properties.[10] One of the first theories to describe filler networks was the "occluded-rubber" theory that stated that when a three dimensional filler network of interconnected particles is formed polymer will be trapped between these particles. This polymer will not be subjected to any external force as it does not interact with the rest of the polymer matrix. The apparent ratio of filler to matrix and in consequence the apparent filler content is enhanced. With increasing strain the filler network is broken and the trapped chains rejoin the matrix. The apparent filler content is decreased and the modulus decreases as well.[57,64]

This concept has been developed further to include explanations for additional experimental observations. The dynamic network model is based on the idea that a network is formed and the modulus of the composite is proportional to the number of filler–filler contacts. These contacts are under external strain continuously broken and reformed. The rate of reformation reduces with increasing strain resulting in a decreasing storage modulus at higher strains.[57]

Polymer–filler interactions

In order to understand polymer–filler-interactions, one has to take the different length scales in a polymer-nanocomposite into account: Particles are usually sized between 1 nm and 100 nm. A typical radius of gyration for a polymer chain is also around a couple of 10 nm up to about 100 nm.[57,78] For a simple nanocomposite with spherical particles and an equal distribution of particles in the matrix the average distance of particles Δ in the matrix can be calculated as

$$\Delta = D \cdot \left(\sqrt[3]{\frac{\pi}{6 \cdot \Phi}} - 1 \right) = D \cdot \left(\sqrt[3]{\frac{\pi}{6} \cdot \frac{m_\Phi + m_p \cdot \frac{\rho_\Phi}{\rho_P}}{m_\Phi}} - 1 \right), \quad (3.13)$$

where Φ is the filler volume fraction, m_Φ the filler content, m_p the filler content, D the diameter of a particle or an aggregate of particles, and ρ_Φ and ρ_P are the densities of the particles and the polymer, respectively.[57] Typical filler contents of about 20 wt.-% to 40 wt.-% and aggregate sizes of up to 1 µm result in average distances between

20 nm and 2 µm. Consequently, the distance between particles is in the same order of magnitude as the radius of one or very few polymer coils combined. Bridging of particles *via* adsorbed polymer chains is therefore easily imaginable. The better the interaction between polymer and filler the stronger this effect will be and the more distinct will be the filler network.[57,79]

Different theories have been developed so far to explain the influence of filler–polymer interactions on the reinforcement. They are all based on the concept that polymer chains interact with the fillers surface a small area around each filler particle. They can interact mechanically stable, e.g. when the chains are bound to the surface *via* a chemical bond, or unstable, if the chains are bound *via* physisorption. Mechanically unstable interactions can be broken under higher external forces but may be reformed.[57,80] Their strength highly depends on the polarity of both the filler and the polymer. If fillers are polar and hydrophilic while the polymer is non-polar or only slightly polar, polymer–filler-interactions are unfavored and phase separation is expected.[1,11,81]

If the polymer and the filler surface do interact one theory is the occurrence of an immobilization effect. The flexibility of the chains in immediate proximity to the surface is reduced drastically and they are in a glassy state rather than in a melt-like state and therefore the modulus increases locally which results in an overall increase of the average modulus of the polymer.[82] This is known as "bound rubber" theory. It is one of the most quoted explanations for polymer–filler-interactions. However, an increase of the glass transition temperature would be expected when a glassy layer is formed around the particles. This should be observable either as an increase in the overall glass transition temperature or through the appearance of a second glass transition at higher temperatures caused by the glassy bound layer of polymer. Nonetheless, this has not been shown experimentally so far.[57]

Similar to the aforementioned dynamic network model, the dynamic adhesion model states that the modulus is proportional to the number of polymer–filler contacts. It takes mechanically stable and unstable contacts into account. Mechanically stable contacts are strain-independent and always lead to a mechanical reinforcement. Mechanically unstable contacts are dynamically broken and reformed. The rate of breaking and formation is strain dependent and explains therefore the decrease in storage modulus with increasing strain.[57,80]

Polymer–polymer interactions

Van-der-Waals-interactions always play a role in polymer physics. However, regarding external stresses in polymeric materials they only play a minor role and are rather to be taken into account when comparing two chemically different polymers. In this case however, the polymer–polymer-interactions that are the most impactful are cross-links. Typically two types of cross-links are to be differentiated: chemical cross-links and physical cross-links, like entanglements.[57,83] In this book only thermoplastic polymers are studied so chemical cross-links will be neglected in the further discussion.

The contribution of polymer–polymer-interactions is strongly dependent on the molecular weight of the polymer. At low molecular masses entanglements are not present. In this regime the viscosity of the polymer increases linearly with the molecular mass:

$$\eta \propto M. \tag{3.14}$$

If the molecular mass of the polymer is higher than the entanglement molecular weight ($M_{n,E}$), which is a substance specific quantity, an exponential dependency can be found:

$$\eta \propto M^{\alpha}. \tag{3.15}$$

For linear polymers, as used within this book, the exponent equals $\alpha = 3.4$ and is independent of the monomer. This is due to the fact that longer polymer chains entangle with each other which restricts viscous flow.[57]

As presented in this chapter, many different aspects have to be taken into consideration when discussing the mechanism behind particle reinforcement of polymers. The concepts that have been presented were all developed to describe empirical results obtained for filled elastomers. No theory has been able to explain all reported observations all at once. When adapting these theories to the composites studied in this book, two major aspects have to be taken into account: As discussed before, only thermoplastic polymers are used. Chemical cross-links do therefore not play any role. The impact of polymer–polymer interactions is therefore expected to be less pronounced. The other aspect is that all the theories assume spherical particles. MMT (section 3.3) has a completely different geometry and has only one dimension in the nanometer-scale. The basic idea of having different contributions to the reinforcement does however still apply.

3.3. Montmorillonite

MMT is a naturally occurring phyllosilicate that forms plate-like crystals which are stacked in a multi layer structure.[33,84,85] In nature, MMT is mostly found in Bentonite that additionally contains some amounts of crystalline quartz, cristobalite, and feldspar.[86] Montmorillonite is a 2:1 smectite clay where each individual layer consists therefore of a central octahedral sheet of aluminum in sixfold coordination with oxygen which is sandwiched by two tetrahedral sheets of silicon-oxygen.[84] Isomorphic substitution of Al^{3+} with e.g. Mg^{2+} results in an overall negatively charged layer. For charge balance, between two layers of MMT there are sodium cations that can be substituted by other alkali- or earth alkali metal cations, most commonly lithium or calcium.[87] Montmorillonite is highly hygroscopic and in

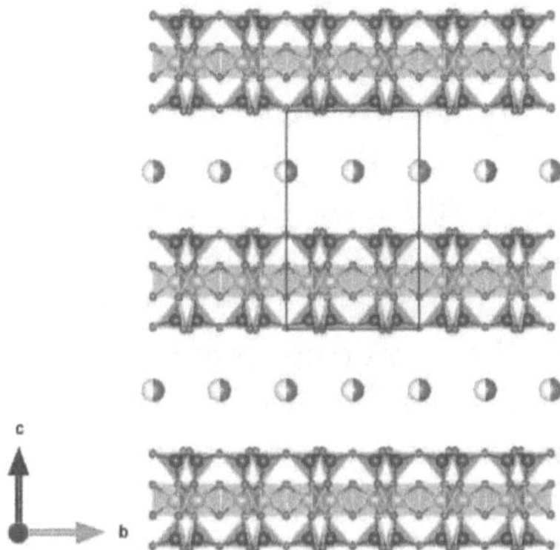

Figure 3.4.: Crystal structure of Montmorillonite along the b-axis. The crystal system is monoclinic, the space group is $C2/m$. The unit cell parameters are $a = 5.18\,\text{Å}, b = 8.98\,\text{Å}, c = 15.00\,\text{Å}$.[88] The real lattice constants can vary depending on the origin of the used MMT. The figure has been made using Vesta (Koichi Momma, Tsukuba, Japan) using data from *Viani et al.*[88]

the swollen state the interlayer cations can easily be exchanged.[39,70,87] The overall formula for one monoclinic repeating unit results to $(\text{Na, Ca, Li})_{0.33}(\text{Al, Mg})_2(\text{Si}_4\text{O}_{10})(\text{OH})_2 \cdot n\,\text{H}_2\text{O}$.[86] The crystal structure is shown in Figure 3.4.

Due to its hydrophilicity,[35,89] MMT is often used as binding material used as seal against water penetration or to increase a solution's viscosity, used for example in drilling muds.[90] It is used as sorbent in nonionic, anionic, and cationic dyes, as well as a catalyst in organic synthesis.[86] Clays in general and Montmorillonite specifically have gained broad attention as inorganic filler material due to their inexpensiveness,[36,91] non-toxicity,[92] chemical[27] and thermal[27,35] stability, flame retardancy,[86] and good mechanical properties.[35,86,93] The first study on the interaction between clay and macromolecules was published by *Bower et al.* in 1949, who observed the absorption of DNA by Montmorillonite.[94] Soon after, different approaches to use the interaction of Montmorillonite and ionic or highly polar macromolecular compounds were published. For instance, in 1960 *Uskov et al.*[94] showed that the addition of MMT modified with octadecylammonium could raise the softening point of poly(methyl methacrylate) significantly. A year later *Nahin* and *Backlund* reported the incorporation of organoclay into a thermoplastic polyolefin matrix showing that the composite materials' tensile strength could be significantly improved.[94] But it was not until researchers of the Toyota group started systematic research on the synthesis and characterization of nylon-6-clay–nanocomposites that this topic became more and more prominent in international research. Toyota reported the synthesis of composites with superior strength, modulus, heat distortion temperature, and water and gas barrier properties.[29–31] They went on to also produce composites of clay and polystyrene, acrylic polymers, rubber, and polyimides.[94] Based on this research, a variety of applications has been developed in the upcoming decades. Montmorillonite–polymer-composites have since been shown to have applications in numerous fields of polymer research. For elastomers and rubbers, historically filled with carbon black, Montmorillonite has been investigated as an additive in order to enhance the abrasion resistance of for example tires.[95] Furthermore, there have been approaches to use Montmorillonite to produce biomimetic nacre-like materials.[13,14,36,96,97]

Chapter 4

Surface Modification of Montmorillonite[*]

Contents

---

[*]The results that are reported in this section are partially reported in "Tuning the Mechanical Properties of Poly(Methyl Acrylate) via Surface-Functionalized Montmorillonite Nanosheets" (*Macromolecular Materials and Engineering, accpeted* 23.10.2020). Parts of this chapter are also part of the bachelor theses "Synthese und Characterisierung von Polymer–Schichtsilikat-Nanokompositen" by Katharina Maria Thien (Georg-August-Univeristät Göttingen, **2018**). and "Synthese von Polymer–Montmorillonite-Filmen: Strategien zur Modifikation der mechanischen Eigenschaften" by Jytte Möckelmann (Georg-August-Univeristät Göttingen, **2019**).

In its pristine form MMT is highly hydrophilic and therefore only miscible with hydrophilic polymers, like for example poly(vinyl alcohol), without severe aggregation of the filler.[14,51,69,91,98] In order to synthesize composite materials of MMT and hydrophobic polymers, different strategies have been developed. Most of them rely on a surface modification of the MMT in order to increase its hydrophobicity.[39,51,69,99] A common approach for the surface modification of for example silica nanoparticles is the anchoring of initiators, monomers, control agents, or polymers onto the surface *via* a stable siloxy bond for subsequent polymerization on the surface.[100] This approach is based on the reaction of hydroxyl groups on the surface with silyl ether groups of the respective reagent. Theoretically, MMT does only bear hydroxyl groups at defects in the crystal structure, as can be seen from Figure 3.4.[101] Additionally, they can be found at the borders of platelets. They can be used for a silylation reaction resulting in covalently bound surface modifying agents. This approach allows for example the grafting of RAFT agents in order to grow polymer brushes on the surface.[100,102] For spherical nanoparticles, as for example silica, this approach leads to very good results. Different grafting densities, different brush lengths, or even bimodal brushes can be grown.[10,81,103] This approach brings the advantage of covalently bond polymer brushes which is a very stable anchoring that is also expected to be temperature stable.[101] For MMT, this has, however, the disadvantage that the polymer chains are mostly anchored at the borders of the MMT platelets due to the availability of OH-groups. The chains will therefore not get in between the layers and will in consequence not lead to exfoliation.

Another more promising approach for the surface modification of MMT is based on ionic interactions. As explained in section 3.3, the individual MMT layers are charged negatively and for charge balance sodium cations are intercalated in the interlayer gallery. These cations can be exchanged for other cations. Typical examples are amino acids, phosphonium salts, or alkylammonium quarternary salts.[32,84,104–106] As these cations replace the intercalated sodium cations between the layers a polymer functionalization *via* these reagents results in intercalated polymer. As the polymer chains are big compared to the thickness of individual MMT layers and can additionally interact with surrounding polymer, exfoliation is furthered. An approach like this is used within this book.

Within this chapter the surface modification as used within this book is presented. Therefore, the synthesis of the surface bound monomer is presented as well as its anchoring to the MMT. Subsequently, the modification with polymer in a *grafting through* approach is presented and discussed. Both, the monomer and polymer modified p articles a re c haracterized w ith s everal m ethods i n o rder to develop a pool of characterization techniques from which can be selected for future projects.

4.1. Premodification of MMT with a cationic monomer

For the preparation of PLSNs an *in situ* two step approach has been chosen. In a first step, a suitable monomer should be anchored to the MMT surface *via* ionic interactions. This should then be used in a *grafting through* approach in a subsequent polymerization.

Salem et al.[39] have presented a strategy to anchor a styrene derivate that is a quarternary ammonium salt to the surface of MMT. They reported the successful *grafting through* polymerization with styrene and methymethacrylate as comonomers. Based on this, *N,N*-dimethyl-*n*-hexadecyl-(4-vinylbenzyl)ammonium-chlorid (VB16) was chosen as surface-anchored monomer (Scheme 4.1). VB16 was synthesized following a literature procedure published by *Salem et al*[39] by addi-

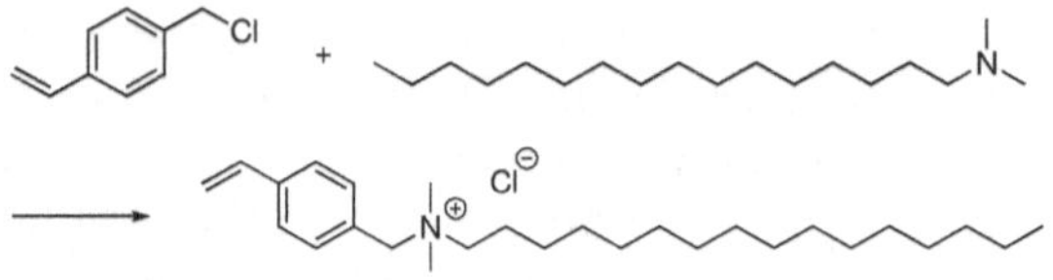

Scheme 4.1: Reaction scheme for the synthesis of VB16 from vinylbezyl chloride and *N,N*-dimethyl-*n*-hexadecylamine. The reaction was carried out in ethyl acetate at 40 °C over night. The resulting white solid was purified by recrystallization from ethyl acetate.

tion of 4-vinylbenzyl chloride to *N,N*-dimethyl-*n*-hexadecylamine in ethyl acetate. After stirring overnight, followed by recrystallization, VB16 was obtained as a white solid (Scheme 4.1).

Prior to modification with VB16 the MMT was allowed to swell in water over the course of one week. This allows water to diffuse in between the layers where a complexation of the interlayer cations occurs. This results in a swelling of the MMT and therefore in an increased interlayer distance that enables larger cations - like the VB16 - to diffuse in between the layers.[86] After swelling, a solution of VB16 in water was added dropwise at 0 °C. After filtration and purification the monomer functionalized MMT (VB16 modified MMT (MMT-VB16)) was received. The process is shown in Scheme 4.2.

4.1.1. Characterization *via* TGA

To confirm the successful surface modification with VB16, TGA was performed. The curve for the unmodified MMT shows an initial step close to 100 °C which can be attributed to the evaporation of water intercalated into the highly hydrophilic MMT. A second step between 550 °C and 725 °C originates from the formation of Anhydromontmorillonite. During this process the crystal structure changes and formerly bound hydroxide ions are excluded. This process is irreversible.[40,107,108] The curve of the MMT-VB16 shows several steps.

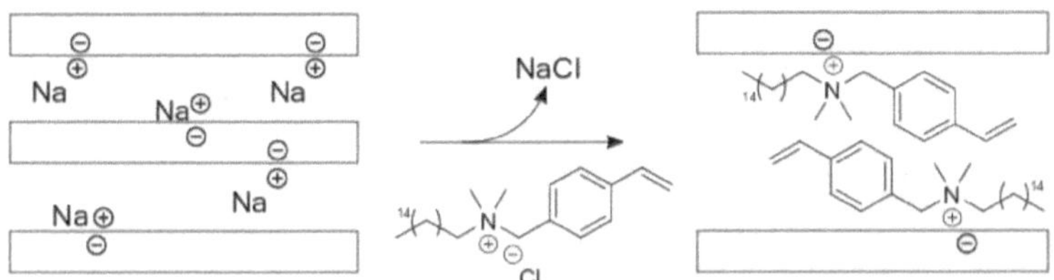

Scheme 4.2: Schematic functionalization procedure. The naturally surface bound sodium cations are exchanged by the organic cation VB16. Sodium chloride is removed by washing diligently with water. Due to the sterically more demanding VB16 the interlayer distance increases.

The step at about $100\,°C$ is significantly reduced. The functionalized MMT was dried *in vacuo* prior to analysis to remove all organic solvent of the functionalization process. This also reduces the water content. The most prominent step at $200\,°C$ can be attributed to the decay of the aliphatic part of the VB16. The aromatic part decays at about $300\,°C$. Furthermore, a step at $600\,°C$ can again be associated with the formation of Anhydromontmorillonite. From TGA, it can be qualitatively concluded that the surface modification of MMT with VB16 was successful.

Evaluation of the thermogram of the pure MMT yields a mass loss of $\Delta m_{MMT} = 6.6\,$wt.-% for the pure MMT. From the used mass for the analysis of MMT-VB16 and the remaining mass at $1\,000\,°C$ the fraction of VB16 in MMT-VB16 can be determined to be $30.4\,$wt.-%.

4.1.2. Characterization *via* EA

Additionally to TGA, quantitative EA was performed. This method allows the determination of the carbon, hydrogen and nitrogen content. The results for MMT and MMT-VB16 are shown in Table 4.1. It can clearly be seen that, contrary to the unmodified MMT, the

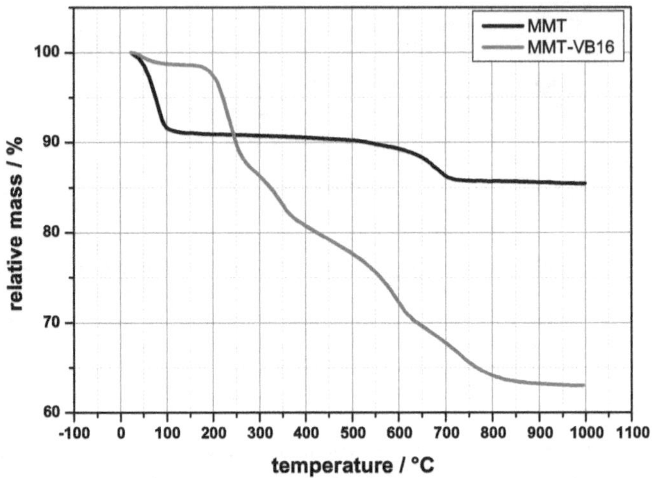

Figure 4.1.: Thermogram for pure MMT and MMT-VB16. Depicted is the relative mass in dependency of the temperature. The samples were heated with $10\,\mathrm{K\,min^{-1}}$ from room temperature to $1\,000\,°\mathrm{C}$.

Table 4.1.: Results of the EA of MMT and MMT-VB16.

sample	carbon /%	hydrogen /%	nitrogen /%
MMT	<0.5	1.54	-
MMT-VB16	25.89	4.22	1.14

functionalized MMT-VB16 shows an organic proportion. The results are in very good agreement with the TGA results as both methods show an organic proportion of about 30 wt.-% within the respective error margins.

4.1.3. Characterization *via* IR-spectroscopy

Another method to confirm the successful functionalization of the MMT with VB16 is IR-spectroscopy. Firstly, ATR-FTIR spectroscopy was performed. Spectra of pure MMT, MMT-VB16, and VB16 were recorded and compared (Figure 4.2).

The spectrum of MMT (purple) was used as a reference. It shows two characteristic bands that can later be used to analyze its composites. One distinct small peak at $3\,624\,cm^{-1}$ corresponding to $O-H$ stretch vibrations of both adsorbed OH-groups and OH-groups in the octahedral sheet of the crystal lattice. At $1\,000\,cm^{-1}$ a band represents the $Si-O$ stretch vibration.[109] The VB16 (yellow) shows a double peak at $2\,844\,cm^{-1}$ and $2\,921\,cm^{-1}$, respectively. They can be attributed to $C-H$ vibrations of the aliphatic part of the VB16. By comparing both curves to the spectrum obtained for the MMT-VB16 (orange), it can be seen that this spectrum contains bonds both characteristic for the VB16 and the MMT. It can therefore be concluded that the functionalization of the MMT with VB16 was successful.

4.1.4. Characterization *via* SAXS

In subsection 4.1.1 to subsection 4.1.3 it was shown that the MMT can be successfully modified with VB16. TGA, EA, and ATR-IR, however, do not answer the question whether the MMT-multilayer stacks are only functionalized on the outside with VB16 or if the VB16 is also intercalated between the individual layers. In order to further validate the surface modification of MMT with VB16, SAXS measurements were performed. Therefore 5 wt.-% of MMT and MMT-VB16, respectively, were disposed onto a PMA support which was also measured without particles in order to perform background correction.

Rawdata and corrected data are shown in Figure 4.3. The scattering curve of PMA shows two peaks. The first peak at a scattering

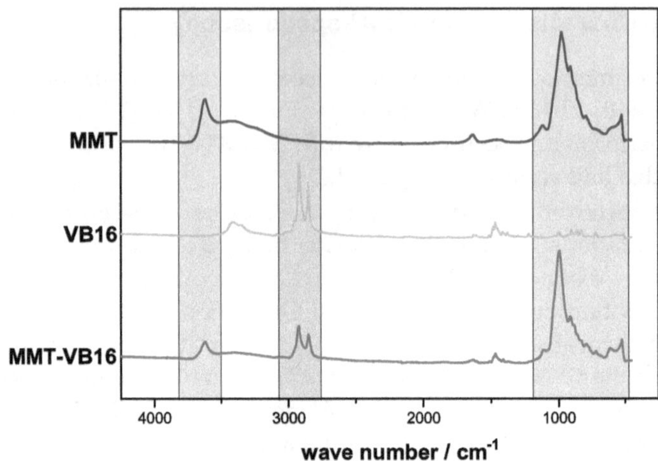

Figure 4.2.: Relative ATR-IR absorbance of MMT, MMT-VB16 and VB16. For the sake of clarity, the curves of MMT and MMT-VB16 are shifted vertically. Highlighted areas represent the characteristic bands for either MMT (blue) or VB16 (grey).

angle of 5.2° originates from a capton foil that is used within the experimental setup. It is therefore observed within any measurement. The broad peak starting at about 6° that is not completely within the measurement range is also observed within any measured sample but is pronounced the most for the pure PMA. This is an amorphous halo peak that is observed for any amorphous material. The more ordered a system is the sharper are the observed scattering peaks. For crystalline materials this results in sharp peaks, whereas for amorphous systems that have only a short range order a very broad halo is observed. The scattering curve for the MMT on a PMA support shows, in addition to the amorphous halo, one distinct peak at about 5.2°. In accordance with Equation 3.10 this corresponds to a d-spacing of 1.68 nm. It has to be noted that this is not the interlayer distance but the thickness of one layer consisting of one monolayer MMT and one times the distance between the two layers, or in other words the dimension of one unit cell which is also depicted in Figure 3.4. The scattering curve of the VB16 functionalized MMT shows four distinct peaks. The first at $2\Theta = 2.5°$ corresponds to a d-spacing of 3.52 nm displaying an increase in unit cell size along the c-axis of 1.84 nm which can be attributed to the intercalation of the sterically more demanding VB16-cation. The additionally appearing peaks are due to scattering of higher orders and correspond therefore to the same interlayer distance (see eq. 3.10).

SAXS measurements confirm the successful increase of interlayer distance due to the intercalation of a cationic monomer.

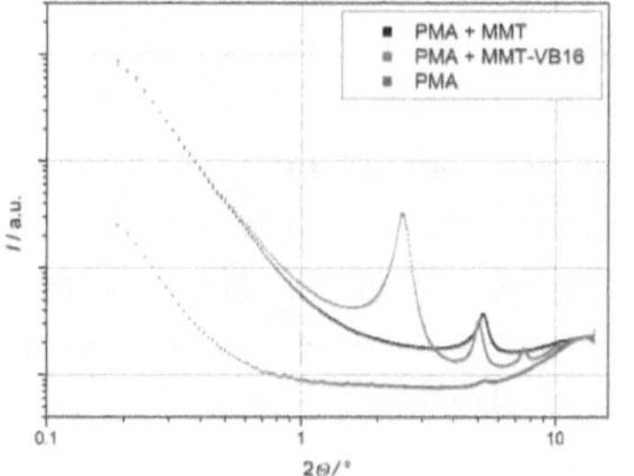

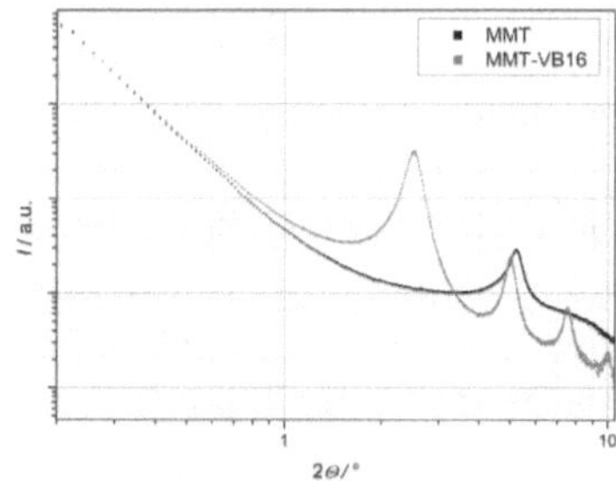

Figure 4.3.: SAXS spectra of MMT and MMT-VB16 particles. Left: Scattering intensity in dependence of the scattering angle 2Θ of MMT, VB16-functionalized MMT and PMA are shown. Both particle types were measured in a PMA matrix. Right: Corrected SAXS spectra.

4.2. Preparation of polymer–MMT-nanocomposites

As comonomer to the VB16, methyl acrylate (MA, Scheme 4.3) was chosen because its polymer, PMA, exhibits a glass transition temperature of about 10 °C. In literature examples of PLSNs with various polymers can be found. Examples include, but are not limited

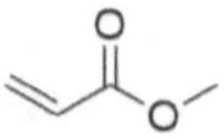

Scheme 4.3: Structural formula of methyl acrylate.

to nylon,[29] poly(styrene),[110] epoxy resins,[98,111] poly(vinyl alcohol), poly(propylene),[112] poly(vinyl chloride),[25] acrylonitrile butadiene styrene copolymer,[113] poly(acrylo nitrile),[114] poly(carbonate),[115] poly (ethylene oxide),[116] poly(imide)[117] or poly(vinyl pyridine).[118] Most of these polymers are below their glass transition temperature at room temperature making them hard and rigid. The influence of the nanosheets on the mechanical properties is however best seen with a polymer that is soft and ductile. PMA is a well known and characterized polymer with properties suitable for analysis of the particles influence. Additionally, it is easily handable with the experimental set ups that were available. It allows the easy formation of testing samples for tensile testing and DMA. A glass transition above room temperature often leads to the formation of inhomogeneous specimens, while a glass transition significantly below room temperature results in specimens which are very soft and already flow at room temperature so that they can not be loaded into the instruments without severe deformation.

In order to prepare the polymer–MMT-nanocomposites, a *grafting through* polymerization was performed. MMT-VB16 was dispersed in toluene. Afterwards, the respective RAFT agent, AIBN as initiator, and monomer were added and the polymerization was performed. As the ratio of monomer to surface bound VB16 was kept higher than the ratio of monomer to RAFT agent, statistically not every chain will bear a VB16 monomer. Therefore, additionally to surface bound polymer also free polymer was formed. The free polymer can be separated from the particles and the surface bound polymer *via* centrifugation. Polymer modified MMT will be named by adding the name of the respective polymer (e.g. PMA modified

MMT (MMT-PMA)). This does refer to polymer modified nanosheets without any free polymer.

4.2.1. Characterization *via* SEC

The characterization of polymers with ionic groups *via* size exclusion chromatography (SEC) is challenging because of interactions between ionic groups and the column material which results in prolonged eluation times.[Staudt2019] Because this method is based on the correlation between eluation time and molar mass, no reliable molar mass distributions of VB16-containing polymers could be obtained with the available setups in our laboratory. Analysis of the free polymer that is formed during the polymerization is a good approximation for the surface-bound polymer as it is known that the reaction conditions in solution and at the surface are sufficiently similar.[39]

Initially, SEC revealed that multimodal distributions were observed (Figure 4.4). This may indicate cross-linking during the polymerization. It is noticeable that the individual maxima are at multiples of the first maximum. It is literature known that through radical transfer to the polymer coupling of multiple species within the polymerization is possible. If a VB16 is included into the growing chain, the benzylic proton of the VB16 is more easily abstractable than the protons from the MA. Thus, a mid-chain radical can form that will then add further monomer and will at some point add to another RAFT group resulting in a species of doubled molecular weight with two (or more) RAFT groups. SEC curves reveal that this occurs for a small percentage of chains. The fact that this multimodal structure is observed for the free polymer indicates that the VB16 is detached from the surface during the RAFT polymerization. Initially, methanol was chosen as solvent for the polymerization as it is known that polar solvents are favored for swelling the MMT.[119–121] The solvent was then changed to toluene. By doing so, the proportion of detached VB16 could be reduced significantly. The respective curves are shown

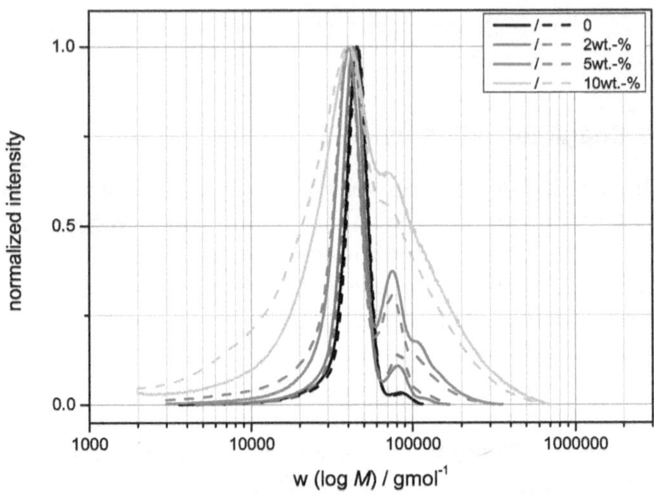

Figure 4.4.: SEC traces of free PMA of a series of polymerizations in the presence of 0, 2, 5, or 10 wt.-%VB16 modified MMT in Methanol. Solid lines are data obtained from RI detection, dashed lines are obtained from UV detection.

in Figure 4.5. The additional maxima are still observable but weaker pronounced. The dispersity $Đ$ is in the range of 1.1 - 1.5. It can therefore be concluded that the polymerization of MA in the presence of monomer functionalized MMT can be performed in a controlled fashion. The best approximation available for characterization of the polymer is therefore the analysis of the free polymer.

It can be found in literature that the surface bound chains can be detached from the surface *via* reversed ion exchange by adding an excess of lithium bromide or lithium chloride. This was performed for the MMT-PMA samples. However, the obtained polymer could not be analyzed with the present SEC setup, as the ionic groups interact with the column material and prevent analysis. It is literature known that the reaction conditions in solution and on the surface

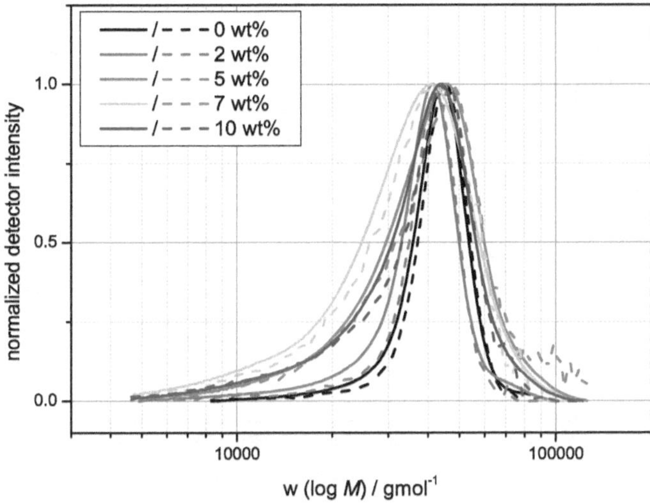

Figure 4.5.: SEC traces of free PMA of a series of polymerizations in the presence of 0, 2, 5 or 10 wt.-%VB16 modified MMT in Toluene. Solid lines are data obtained from RI detection, dashed lines are obtained from UV detection.

are somewhat similar and the molar weight distributions of surface-tethered and free polymer are anticipated to be comparable.[39]

4.2.2. Characterization *via* TGA

Additionally, TGA was performed. In addition to the curves of MMT and MMT-VB16 (black and blue, respectively), as shown before, Figure 4.6 shows curves for the polymer functionalized MMT. The yellow curve is the curve obtained from a measurement of the nanocomposite as-received, meaning a composite from PMA and MMT-PMA. Any additional mass loss compared to the curve for MMT-VB16 can be attributed to PMA. From this curve the MMT content of the nanocomposite can be determined. In order to evaluate

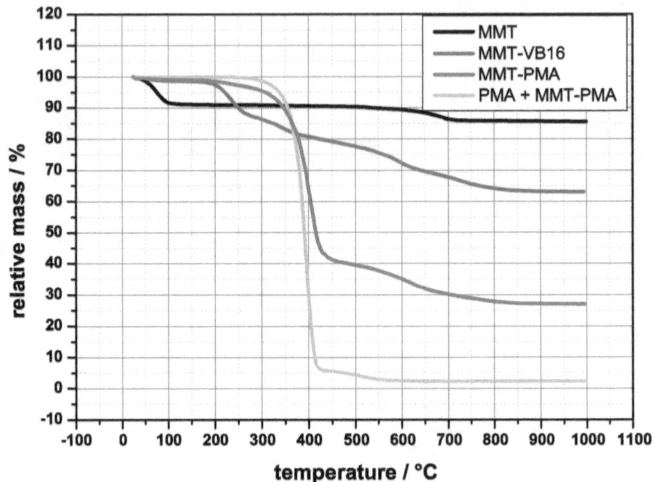

Figure 4.6.: Thermogram of polymer functionalized MMT and the respective nanocomposites. For comparison, the curves for unmodified MMT and monomer functionalized MMT (MMT-VB16) are shown again.

if the polymer was successfully bound to the surface, the nanocomposite was dispersed in dichloromethane (DCM), which is a good solvent for PMA, and then centrifugated. The supernatant was discarded and the process was repeated. Hereby, any free polymer was washed away and again TGA was performed. As it can be seen in Figure 4.6, the mass loss is lower than for the nanocomposite of particles and matrix, showing that there was indeed free polymer, but is significantly higher than for the MMT-VB16. It can therefore be concluded that a significant amount of polymer is bound to the surface of the MMT platelets.

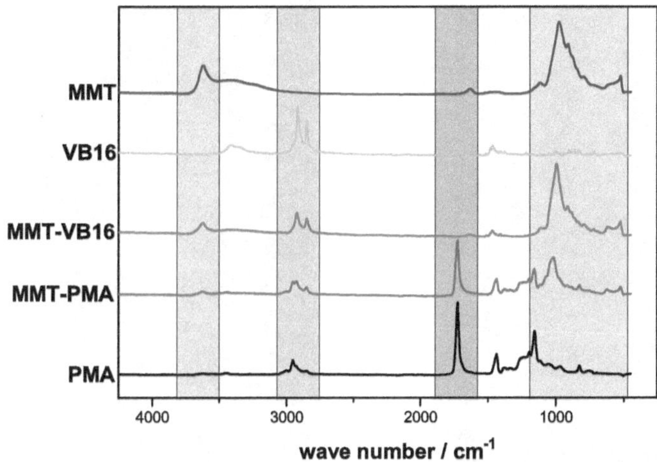

Figure 4.7.: Relative ATR-IR absorbance of MMT, MMT-VB16, VB16, PMA, and MMT-PMA. For the sake of clarity, curves are shifted vertically. The blue highlighted areas represent the characteristic bands for either MMT, grey for VB16, and red for PMA.

4.2.3. Characterization *via* IR-spectroscopy

Figure 4.7 shows, in addition to the beforehand shown spectra of MMT, VB16, and MMT-VB16, two spectra of PMA and MMT-PMA. It can be seen that the characteristic peak at about $1\,700\,\text{cm}^{-1}$ that can be attributed to the carbonyl-group of the acrylate is present in the MMT-PMA sample. ATR-FTIR does therefore support the TGA results as the characteristic bands of the respective educts are also observable in the composite materials.

Additionally, the combination of atomic force microscopy (AFM) and IR-spectroscopy gives further insight. From ATR-FTIR it is only proven that all of the expected components are in the sample but not known where exactly they are located. The samples were diligently washed with DCM prior to investigation, so it is in principle expected

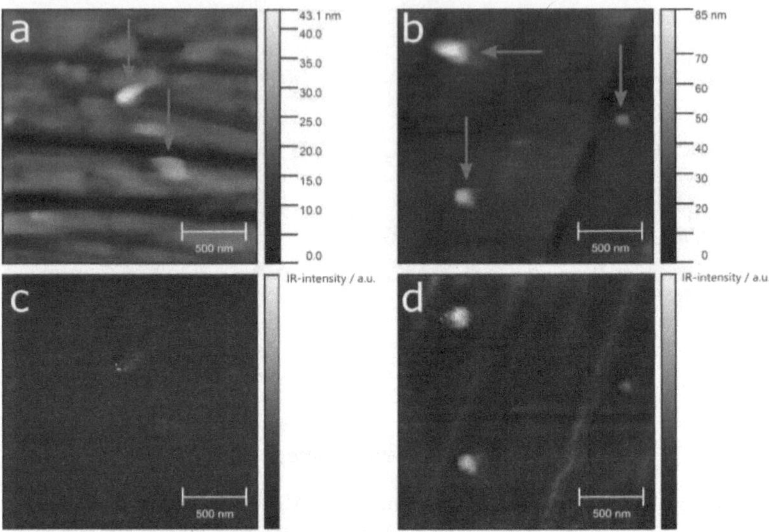

Figure 4.8.: s-SNOM images of single MMT sheets (left) and MMT-PMA (right). Shown are the topology maps (top, a and b) and the IR intensity for the C=O band at 1680 cm-1 (bottom, c and d). Red arrows in the topology graphs indicate individual MMT particles.

that any free polymer is washed away and that the signals originate from surface-grafted polymer only. A strategy to show that for MMT-PMA samples polymer is only located around the particles is the combination of imaging with nm-spatially-resolved infrared spectroscopy, which is realized by the scanning scattering near-field optical microscopy (s-SNOM)-technique. Optical NIR imaging is performed by detecting the backscattered light interferometrically. At the same time the topography of the sample is scanned as known from standard AFM measurements. The AFM-tip is illuminated with an infrared laser and a local IR-spectrum with nm-resolution is recorded.[122,123] This measurement is then performed for a sample of pure MMT and a sample of MMT-PMA. This allows to obtain a map of the signal intensity of individual IR-bands, corresponding to the occurrence of specific chemical groups. To analyze the distribution of PMA within the sample, the C=O-band at $1\,680\,\mathrm{cm}^{-1}$ was recorded. The particles were dispersed in propylene glycole monomethyl ether acetate (PGMEA) and drop cast onto calcium fluoride waver that shows no signal of its own in the range observed. The solvent was then allowed to evaporate. Figure 4.8 shows the respective AFM topology images (a and b) and IR band intensity (c and d). For the chosen wave number there is a clear response localized around the PMA-functionalized MMT particles (Figure 4.8 d) and no detected response when analyzing the unmodified MMT (Figure 4.8 c) proving the successful polymer-modification of individual MMT sheets without any remaining free PMA.

4.2.4. Characterization *via* SAXS

TGA revealed that the *grafting through* polymerization of MA with monomer modified MMT was successful. TGA can however not give any information whether the polymer is only attached to the surface of a MMT-stack or also intercalated between the layers or if there is even exfoliation. To answer this question SAXS measurements were performed.

Figure 4.9 shows the scattering curves for MMT and MMT-VB16 which have already been discussed in subsection 4.1.4. Additionally, the scattering curve for a nanocomposite is shown. It can be seen that at about $2\Theta = 2.56°$, corresponding to a d-spacing of approximately 3.52 nm, a broad peak with a low intensity can be detected. Additionally, a peak at about $2\Theta = 0.43°$ can be observed. The sharper a peak is the more perfect is the periodic arrangement of the observed structure. Hence, a broad peak indicates the existence of a periodic structure with either only very few repetitions or very irregular periodicity. For the structure of the composite this means that there is

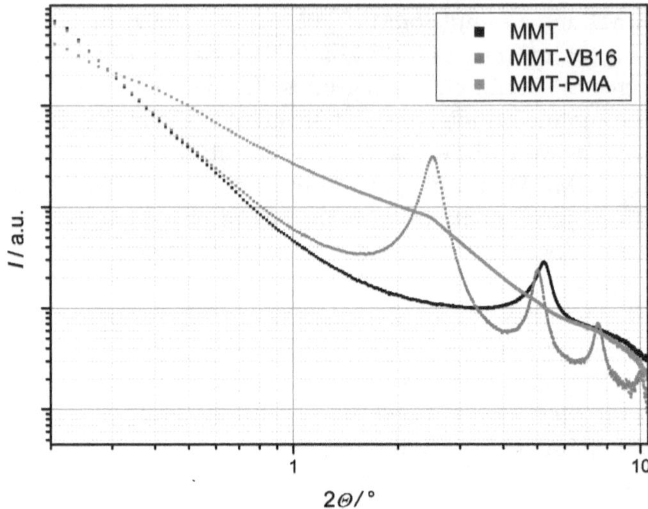

Figure 4.9.: Scattering curves of pure PMA, MMT as well as MMT-VB16 in PMA and MMT-PMA. Shown is the scattering intensity in dependency of the scattering angle 2Θ.

one structure with a very large irregular distance (corresponding to the peak at about $2\Theta = 0.43°$) and another structure where the initial distance has not changed as the peak is still at the same position as for the MMT-VB16 (peak at $2\Theta = 2.56°$) but with significantly lower number of repeating units. The larger scale structure might be explained as a structure where VB16-monomers of different layers are included into the same polymer chain. In this case two MMT layers are bound to the same chain and are therefore still connected but with a larger distance as before. These results indicate that the structure of the MMT in the MMT-PMA either consists of very few periodic units or is generally very irregular. Both interpretations are fully in line with a loss of the periodic layer structure, intercalation of polymer, and partial exfoliation.

4.2.5. Characterization *via* TEM

TEM is a powerful tool to visualize structures on the nanometer scale. Therefore a dispersion of umodified and PMA-modified MMT was drop-cast onto TEM grids. The grids are covered with an amorphous carbon net of an approximate thickness of 4 nm. Figure 4.10 shows TEM images of both samples. From a first view it is apparent that the modification with polymer resulted in a drastic change in structure. The unmodified MMT consists of big chunks of MMT. Towards the edges it can be seen that the structure is thinner. The modified MMT, in contrast, only consists of few layers of MMT as can be seen by comparison to the pure grid. Additionally, at many spots the grid is visible through the MMT layers indicating further that the sample is very thin. The outlines of many individual layers are visible within this part of the sample. SAXS measurements showed before that the polymer was successfully intercalated and that the periodic structure of the MMT was lost. TEM analysis confirms that the reason is, indeed, that most layers are separated from each other.

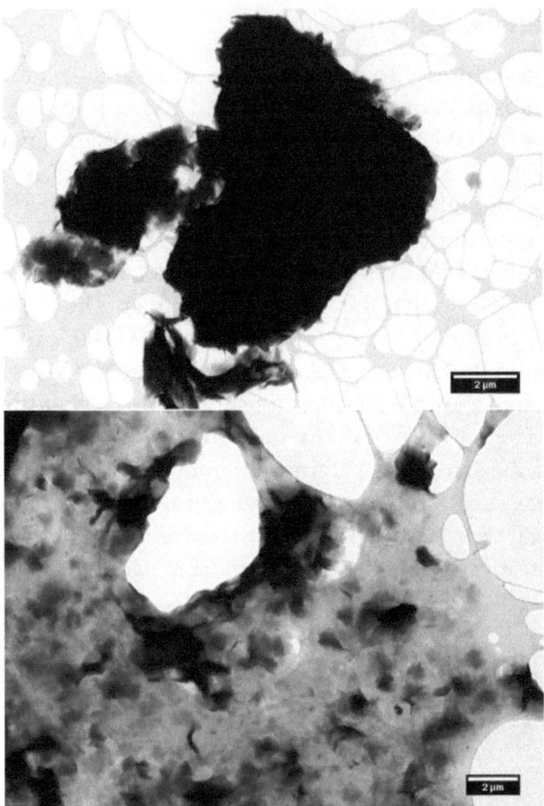

Figure 4.10.: TEM images of unmodified MMT (top) and polymer-modified MMT (bottom).

4.3. Conclusion

Within this chapter a reliable strategy for the surface modification of MMT is presented and the reader is provided with a toolbox of analysis methods to validate the monomer and polymer modification of MMT. It could be shown that the layered silicate can be modified with a cationic monomer *via* an ion exchange procedure up to about 100 % of its literature known ion exchange capacity.[39] The modification with VB16 was confirmed *via* TGA, EA, ATR-FTIR, and SAXS. While the first three methods only provide the information that modification results in a grafted organic proportion, SAXS also reveals that, as was aimed for, the monomer is successfully intercalated between the layers. Polymer modification was carried out *via grafting through* RAFT polymerization of different monomers. SEC revealed that the system is quite sensible to the solvent used. Methanol as a solvent which is commonly used for standard RAFT polymerization of PMA resulted in detachment of some grafted chains from the surface and showed that in small amounts the transfer-to-polymer occurs resulting in multimodal molar mass distributions. Obtained dispersities, however, still confirmed controlled polymerization conditions. Changing the solvent to toluene prevented any detachment of surface bound chains. However, it can be assumed that transfer-to-polymer still occurs. TGA confirmed a drastic increase of the organic proportion after polymerization, further confirming successful polymerization. This was additionally supported by ATR-FTIR measurements that proved the presence of carbonyl groups within the particles that can only originate from the polymer. Furthermore s-SNOM measurements, combining IR spectroscopy and atomic force microscopy, showed that when dispersing the particles in a solvent and dropcasting them onto a surface carbonyl groups and therefore polymer can only be found at the exact location of the particles. SAXS showed that intercalation and partly exfoliation of the MMT was successful. Lastly, TEM gave an optical impression of the change in particles form highly aggregated stacked MMT particles to thin dispersed layers of polymer grafted MMT.

Chapter 5

Preparation and Characterization of MMT-filled Polymer-Nanocomposites[*]

Contents

The modification of a polymer's mechanical properties through

---

[*]The results that are reported in this section are partially reported in "Tuning the Mechanical Properties of Poly(Methyl Acrylate) via Surface-Functionalized Montmorillonite Nanosheets" (*Macromolecular Materials and Engineering, accpeted* 23.10.2020). Parts of this chapter are also part of the bachelor book "Synthese und Charakterisierung von Polymer–Schichtsilikat-Nanokompositen" by Katharina Maria Thien (Georg-August-Univeristät Göttingen, **2018**).

nanoclay is well known. For polymer–clay-nanocomposites various examples can be found in literature. *Chan et al.* prepared nanoclay/epoxy composites with varying filler content.[124] They found that with increasing filler content the *Young's* modulus and tensile strength could be increased. *Wang et al.* prepared MMT/poly(vinyl alcohol) (PVA)-nanocomposites.[125] Their composites covered the whole range of 0 wt.-% to 100 wt.-% filler. They found an enhanced mechanical performance up to a silicate content of 70 wt.-% filler, after which they observed deteriorated mechanical properties. Other examples include for example composites of varying amounts of MMT and poly(methly methacrylate) (PMMA),[1] epoxy,[99] PVA,[91] zein,[126] or polyimide.[27] These studies have in common that they investigated polymers that are below their glass transition temperature at room temperature and are therefore rigid. In this study, the influence of the MMT on the mechanical properties of a polymer above its glass transition is studied. PMA was chosen because its glass transition is below room temperature, but at room temperature it is still rigid enough to form tensile testing specimen. Additionally, its properties are well characterized.[127]

All samples discussed in this chapter consist of a PMA matrix filled with differently surface modified MMT nanosheets.

To analyze the influence of the surface modification on the mechanical properties, different experiments were conducted. First of all, the influence of organic surface modification on the mechanical properties was investigated by preparing nanocomposites containing either unmodified, monomer modified or polymer modified particles at the same particle content. Afterwards, the influence of the particle content for nanocomposites containing polymer modified MMT was investigated. Therefore, a series of samples containing up to 20 wt.-% MMT was prepared. For the first time the influence of a variation of the grafted chain length while keeping the matrix chain length constant was investigated. This further expands the field of PLSNs.

From these experiments, further insight on how particles influence
the mechanical properties of PLSN is expected.

5.1. Influence of the surface modification on the mechanical properties

As explained in section 3.2 the success of the enhancement of the
mechanical properties of a polymer by the addition of nanoparticles
is determined by two major factors: the state of particle dispersion
in the matrix and the interaction of the particles with the matrix.
In order to determine whether the functionalization of MMT with
polymer is a valid strategy to enhance the mechanical properties of
polymers, different nanocomposites of PMA matrix polymer ($\overline{M}_n$ =
$3.6 \cdot 10^4\,\mathrm{g\,mol^{-1}}$, $Đ$ = 1.1) filled with either unmodified MMT, VB16-
modified MMT, or PMA-modified MMT, all at the same overall MMT
content, were prepared and analyzed. The polymer-coated MMT
nanosheets were grafted with PMA of a number average molecular
mass of $\overline{M}_n$ = $3.6 \cdot 10^4\,\mathrm{g\,mol^{-1}}$ ($Đ$ = 1.4). Representative stress–strain-
curves are shown in Figure 5.1. The resulting *Young's* moduli and
values for the tensile strength, strain-at-break and toughness are
given in Table 5.1 and Table 5.2.

Table 5.1.: Calues for *Young's* modulus E, tensile strength σ_y, and stain at break
ε_B determined by tensile testing for a series of nanocomposites with
differently modified MMT. The respective stress–strain-curves are
shown in Figure 5.1.

sample	type of MMT	E/ N mm^{-2}	σ_y/MPa	ε_B/ 10^3 %
A1	none	36.8 ± 1.7	0.6 ± 0.1	1.86 ± 0.18
A2	unmodified	87.6 ± 8.6	1.3 ± 0.1	1.04 ± 0.06
A3	VB16 modified	70.2 ± 9.6	2.0 ± 0.3	1.19 ± 0.19
A4	PMA modified	96.0 ± 7.3	5.3 ± 3.7	0.32 ± 0.11

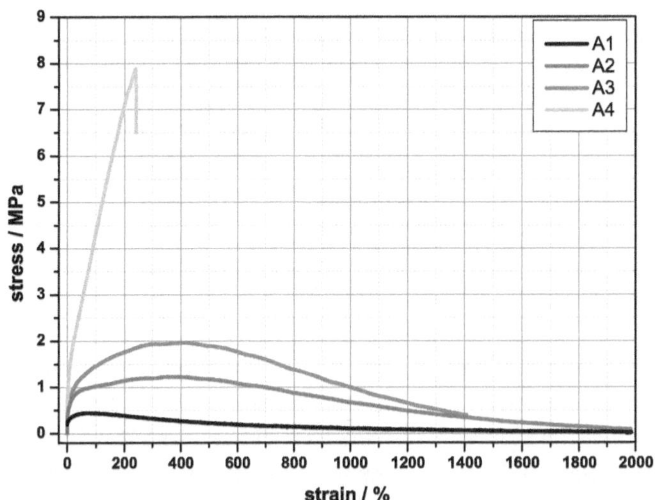

Figure 5.1.: Stress–strain-curves of a series of nanocomposites containing either
no filler (A1), umodified MMT (A2), VB16 modified MMT (A3),
or PMA modified MMT (A4). All samples contain 5 wt.-% MMT.

Both the additions of unmodified MMT or MMT-VB16 (samples
A2 and A3) lead to an increase in *Young's* modulus, tensile strength
and toughness compared to pure PMA (sample A1) whereas the highest reinforcement is observed when PMA modified MMT (sample
A4) is used as filler. Unmodified hydrophilic MMT and hydrophobic PMA are expected to be immiscible and therefore form a phase
separated structure with predetermined breaking points.[81,94] Consequently, they are not promising for the production of composite
materials with superior properties that are tailorable over a wide
range. The highest reinforcement is observed when the filler and
the matrix can interact with each other *via* the surface grafted PMA
resulting in a homogeneous composite material with well dispersed
exfoliated MMT due to the enhanced polymer–filler-interactions. Due

Table 5.2.: Toughness U_T determined by tensile testing for a series of nanocomposites with differently modified MMT. The respective stress–strain-curves are shown in Figure 5.1.

sample	type of MMT	$U_T/\ 10^6\ \mathrm{J\ m^{-3}}$
A1	none	3.6 ± 0.3
A2	unmodified	13.5 ± 1.2
A3	VB16 modified	18.1 ± 2.0
A4	PMA modified	10.1 ± 2.3

to these interactions stress can be properly distributed within the material. Composites using polymer coated nanosheets therefore make a promising strategy to tune the mechanical properties of PMA.

Additionally, DMA was performed to get further information on the mechanical performance. The evaluation of the loss factor $\tan(\delta)$ allows for the analysis of the viscous and elastic properties. Respective temperature dependent measurements of the loss factor for samples A1-A4 are shown in Figure 5.2. All samples show a clear peak around the glass transition. As shown in subsection 3.1.3, at the glass transition the storage modulus drops for up to three orders of magnitude as the material undergoes the transition from glassy to amorphous. At the same time, the loss modulus exhibits a peak. Energy that is brought into the system as strain is used for chain movement. In the beginning of the glass transition region it leads to an increased internal friction resulting in energy loss and observed by an increase in loss modulus. When the temperature rises further, the chains are mobile enough so that the friction is reduced. Consequently, the loss modulus decreases. Combined, this results in peaks in the loss factor. From the definition of the loss factor (Equation 3.9) it is known that for $\tan(\delta) > 1$ the viscous properties dominate the materials properties.

DMA confirms the observations made in tensile testing experi-

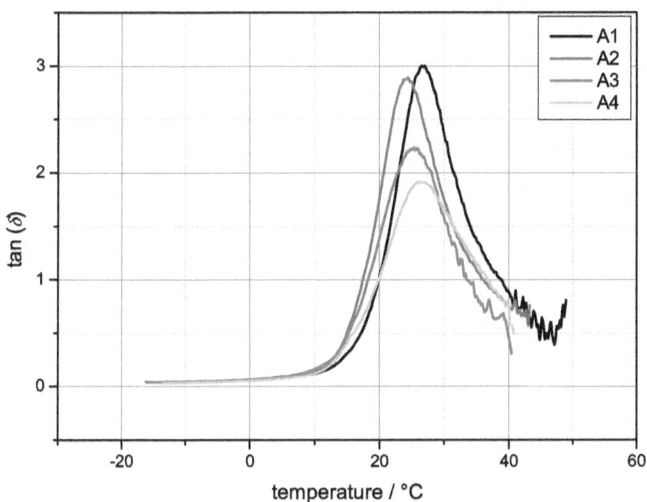

Figure 5.2.: Loss factor $\tan(\delta)$ measured *via* DMA of various samples of
nanocomposites containing particles with different functionalization.

ments. PMA (sample A1) has the highest loss factor, followed by
composites with unmodified MMT and MMT-VB16, while the composite with MMT-PMA (sample A4) exhibits the lowest loss factor.
This is in good agreement with the results from tensile testing which
also showed a decrease in plastic deformation when going from
unmodified MMT as filler to PMA modified MMT. The loss factor of
sample A4 is still larger than one indicating that the viscous properties still outweigh the elastic properties. Comparison to the tensile
testing results confirms this as this sample still shows a strain-at-
break of $(3.2\pm1.1)10^2$ % and can therefore still be classified as a
ductile material despite being stiffer and less ductile than samples
A1-A3.

It can be concluded that the modification of with PMA is a promising strategy for tuning the mechanical properties of PLSNs of PMA and MMT.

5.2. Influence of the MMT content on the mechanical properties

The variation of the filler content is a widely used approach to tune the mechanical properties of polymer-nanocomposites.[19,27,128,129] One of the simplest approaches to explain the effect fillers have on the mechanical properties of a composite is known as *Einstein–Smallwood*-relation[128] and is based solely on the assumption that the modulus of the composite is determined by the modulus of the matrix and the filler content (Equation 3.12) and thus results in a linear dependency of the modulus on the filler content. This relation has been observed for various nanoparticles in a polymer matix.[130–132] It is also known that the geometry of the particles has a strong impact on this effect. While the *Einstein–Smallwood*-relation was introduced for spherical particles, tube or sheet like nanoparticles have been shown to have a different effect on the modulus. *Yudin et al.* showed that, while still being linear, the increase of the modulus with increasing filler content is much stronger, being strongest for the nanosheets.[27]

To analyze the effect of the variation of the MMT content on the mechanical properties, in a first step, polymer modified MMT particles ($\overline{M}_{\text{n, surface}} = 3.4 \cdot 10^4\,\text{g\,mol}^{-1}$, $Đ = 1.3$) were synthesized by the previously presented *grafting through* approach (see section 4.2). The particles and the present free polymer were then separated *via* centrifugation in DCM. The solubility of PMA in DCM is very high, so the solid particles and the surface bound polymer are settled down in the centrifuge tube while the free polymer remains in solution. As the particles are best redispersable when kept in a wet state, they were not dried after centrifugation but redispersed in different concentrations in PGMEA. The matrix polymer ($\overline{M}_{\text{n}} = 3.0 \cdot 10^4\,\text{g\,mol}^{-1}$,

$Đ$ = 1.1) was added immediately. The resulting nanocomposites were then cast into molds and analyzed *via* tensile testing. Figure 5.3 shows exemplary stress–strain-curves for all composites. As can be seen, already the addition of only 5 wt.-% of polymer modified MMT leads to a drastic change in stress-stain performance. The pure matrix polymer is a very soft material that already flows when a very small stress is applied. It can be elongated up to a strain of 2 000 % without breaking. It is well established that amorphous polymers show both elastic and plastic deformation.[57] The elastic deformation originates in two mechanisms: an applied stress results in stretching and distortion of the covalent bonds within the polymer backbone. Additionally, small segments of the polymer can be distorted under external stress. Both these deformations are reversible on the atomic length-scale and result in relatively small deformations on the macroscopic length-scale. Plastic deformation occurs when the polymer is deformed over the point where elastic deformation takes place. Typically, chain sliding, stretching, rotating and entanglement under external load contribute to plastic deformation. This results in chain orientation and constant, non-reversible deformation.[133]

All particle containing samples show a significantly reduced strain-at-break and highly increased yield stresses. Both properties increase further with increasing filler content. For the samples containing 15 wt.-% or more filler no real flowing behaviour is observed as for the samples with lower filler content and especially the matrix polymer. The addition of particles does seemingly reduce the mobility of the polymer chains and restricts their free movement within the sample. Differential scanning calorimetry (DSC) measurements (see Figure 5.4) of samples containing various amounts of MMT-PMA show two glass transition regions. One corresponds to the glass transition of the pure PMA at about 0 °C, as obtained from comparison to the presented DSC curve of PMA. Additionally, a second, less pronounced glass transition at about 38 °C can be observed. For these samples, the molecular weight of the matrix chains and the surface

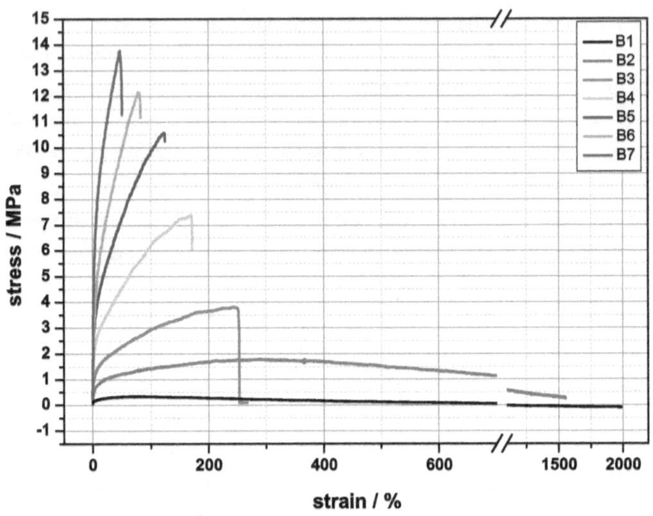

Figure 5.3.: Stress–strain-curves for nanocomposites of matrix polymer and different contents of polymer modified MMT. For filler contents compare 5.3.

bound chains is nearly identical. The chain length dependency of the glass transition is therefore not expected to be the origin of the second glass transition temperature (T_g). A reduced chain mobility, however, results in an elevated glass transition temperature as the chains need more energy to move freely which can be observed here. The higher the particle content, the more pronounced is the second transition. This supports the theory that the particles lead to a reduced mobility of at least parts of the sample.

Additionally, Table 5.3 gives the obtained values for the *Young's* modulus and toughness depending on the filler content. It can be seen that the relation between *Young's* modulus and filler content is nearly linear. Over the course of the series of samples, the *Young's* Modulus could be nearly increased by a factor of 50. The toughness,

Table 5.3.: Values for *Young's* modulus E, tensile strength σ_y, stain at break ε_B, and toughness U_T of nanocomposites with varying MMT-PMA content. Representative stress–strain-curves are shown in Figure 5.3.

Sample	content / wt%	E / 10^2 N mm^{-2}	σ_y / N mm^{-2}	ε_B / 10^2 %	U_T / 10^6 J m^{-3}
B1	0	0.11 ± 0.01	0.3 ± 0.1	18.0 ± 2.4	1.7 ± 0.2
B2	5.4	0.22 ± 0.10	1.9 ± 0.2	15.6 ± 0.9	14.8 ± 1.1
B3	10.9	0.91 ± 0.23	3.7 ± 0.2	2.6 ± 0.4	6.9 ± 2.3
B4	14.9	1.72 ± 0.33	7.9 ± 0.7	1.6 ± 2.0	9.1 ± 0.4
B5	16.5	2.79 ± 0.48	10.9 ± 0.4	1.6 ± 2.2	12.7 ± 2.2
B6	19.9	2.78 ± 0.16	11.8 ± 1.4	0.9 ± 0.3	7.2 ± 0.1
B7	22.6	5.79 ± 0.59	13.5 ± 0.4	0.5 ± 0.1	4.9 ± 0.5

however which is a measure for the energy needed for the sample to break, is more or less constant within the estimated error. The simultaneous decrease in elasticity and increase in yield stress compensate each other.

From what is known about the deformation mechanism of filled polymers and unfilled polymers, energy is needed for different processes. In unfilled polymers or polymers with low filler content which show a significant plastic deformation the energy is needed to disentangle the polymer coils to align them and to pull them past each other. When the polymer is filled with nanoparticles, considerably more energy is needed for elastic deformation due to polymer–filler-interaction and filler–filler-interaction within the filler network. It is known that especially when clusters of rigid filler particles are formed the hydrodynamic effect causes the material to become stiffer overall.[76] At increased filler loading this is expected to occur and can be seen in the obtained characteristics and stress–strain-curves that clearly show a transition from mainly plastic to a mainly elastic, rigid, stiff material. Furthermore, one theory to explain the effect that fillers have on a polymeric matrix states that surface grafted

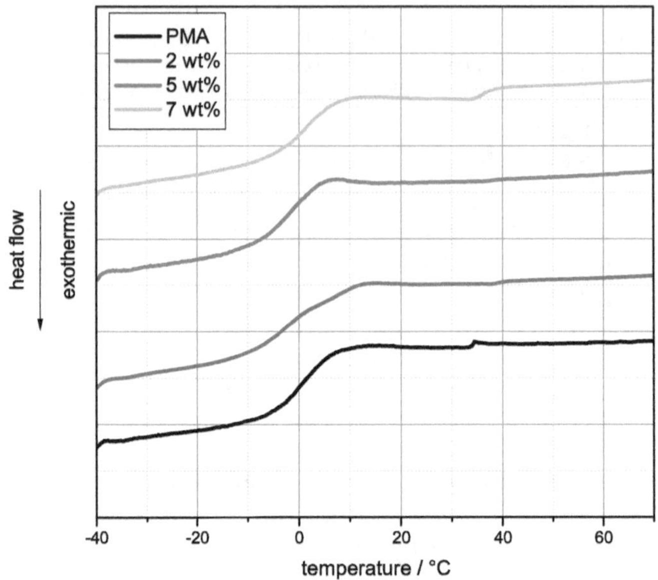

Figure 5.4.: DSC curves of PMA and PMA containing polymer modified MMT. Curves are shifted vertically for visual clarity.

chains have a significantly reduced mobility. Around particles a glassy layer with a considerably enhanced modulus and glass transition temperature is formed as also confirmed by DSC here. Around this glassy layer a second layer of reduced chain mobility is formed. This may extend up to 10 nm. With increasing particle loading the proportion of polymer that is at least partly incorporated into one of those areas significantly rises.[134] An overall reduced chain mobility is the consequence and can be observed perfectly in the presented stress–strain-data.

Additionally, DMA measurements were performed. Figure 5.5 shows the loss factor obtained from DMA measurements of samples

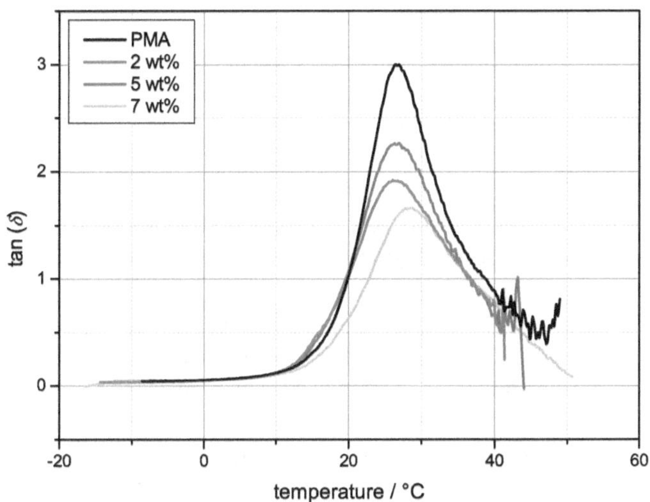

Figure 5.5.: Loss Factor measured *via* DMA of various samples of nanocomposites containing different amounts of PMA modified MMT and PMA without filler.

containing 0 wt.-% to 7 wt.-% of PMA modified MMT. It can be seen that with increasing filler content, the loss factor decreases. This indicates that the material becomes less viscous. However though, the loss factor is still larger than one so the viscous properties are still predominant. These results are in good agreement with the beforehand presented results from the tensile testing which also show a decrease in plastic deformation and therefore less viscous flow with increasing filler content.

Subsequently, creep behaviour of these samples was measured using again DMA. A specimen is loaded with an external stress for a defined period of time. Figure 5.6 (top) shows the applied stress-profile. It is measured how fast the sample reacts to the external stress and how strong the deformation is. The answer of the sample consist of two contributions: an elastic, instantaneous answer and a viscous answer that is changing over time.[135] After the stress is removed the recovery of the sample is detected. Figure 5.6 additionally shows the obtained curves for samples containing up to 7 wt.-% of MMT-PMA. All samples show an initial, immediate displacement corresponding to the elastic part of the mechanical response. It can be seen that the higher the particle content, the lower the initial displacement. This is in good agreement with the observations that the addition of higher amounts of particles results in a stiffer material. It was already shown that the addition of particles leads to an increase in the *Young's* modulus which is a measure for the resistance of a material against elastic deformation. All samples in this experiment are loaded with the same force of $F = 2\,\text{N}$. The sample with the highest modulus does therefore display the smallest displacement. The elastic response is followed by a second, time dependent increase in strain. This corresponds to the viscous answer of the sample. It can be seen that the pure PMA and the sample containing only 2 wt.-% show the smallest increase over time while the samples with 5 and 7 wt.-%, respectively, show a more pronounced increase of strain over time. The higher the deformation over time, the more plastic flow occurs. This is at first glance contra intuitive, as the samples with higher particle loading have already been shown to exhibit less plastic flow. However, the deformation observed in this experiment is way below the strain-at-break observed in a tensile testing experiment and it is not observed how much the material can be deformed but rather how fast it deforms. Comparison to tensile testing is therefore not valid. In order to understand the observed results, the loss modulus E'' of the materials at 28 °C has been measured (Table 5.4). This illustrates that enhanced creep behaviour is a result of the increased loss modulus which is a measure for the share of the energy that is lost through plastic deformation, is significantly increased with increasing filler amount.

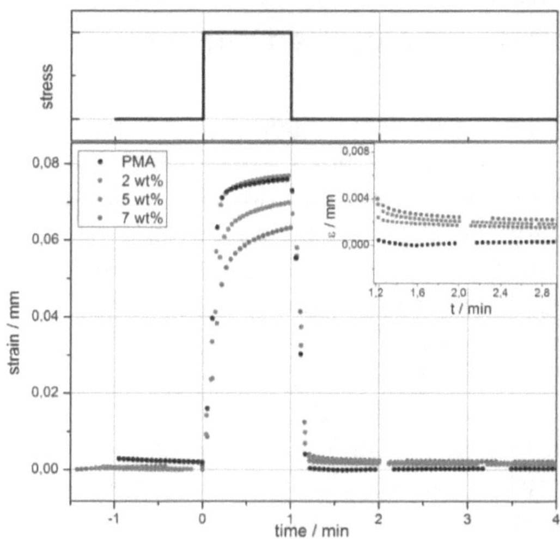

Figure 5.6.: DMA creep measurements of samples containing different amounts of MMT-PMA and stress-profile. Measurements were performed at (27.6 ± 0.7) °C. The external force equaled 2 N.

Table 5.4.: Loss modulus of samples containing different amounts of MMT-PMA.

MMT-PMA content / wt.-%	loss modulus / 10^7 Pa
0	2.1 ± 1.0
2	2.4 ± 1.4
5	5.5 ± 1.1
7	12.2 ± 1.2

5.3. Modification of the mechanical properties by variation of the grafted polymer chain length

In section 5.2 the mechanical properties of samples containing particles whose surface bound polymer chains had always the same molecular weight were analyzed. However, it is easily imaginable that for the polymer–filler-interactions the molecular weight of the surface tethered chains plays a key role. It is well known that very short polymer chains do not entangle with each other. For each polymer a characteristic $M_{n,\mathrm{E}}$ is determinable which is the average molar mass between two entanglements. For PMA this is literature known to be around $1.1 \cdot 10^4\,\mathrm{g\,mol^{-1}}$.[136] It would therefore be expected that this chain length has to be reached to ensure good interaction between the matrix and the surface bound chains. To investigate this further, different nanocomposites with identical matrix ($\overline{M}_\mathrm{n} = 3.0 \cdot 10^4\,\mathrm{g\,mol^{-1}}$, $Đ = 1.1$) and varying surface grafted chain length have been prepared. It was taken into account that with the surface modification approach used, the polymer amount on the particles is increased, when the chain length increases. As it was already shown before, the MMT content has a massive impact on the mechanical properties of composites. The added amount of matrix polymer was therefore carefully adjusted to ensure the same overall MMT content for each sample. *Via* subsequent TGA-measurements it was ensured that the MMT-content was kept constant. All resulting composites were solution cast into moulds and then analyzed *via* tensile testing. The resulting stress–strain-curves are shown in Figure 5.7 and the respective values characterizing the mechanical performance are given in Table 5.5.

As soon as the entanglement molecular weight is surpassed (sample C3 and higher) the general behavior of the sample is always the same: in the beginning, a quick linear rise of the stress with strain is observed. After the maximum the stress drops followed

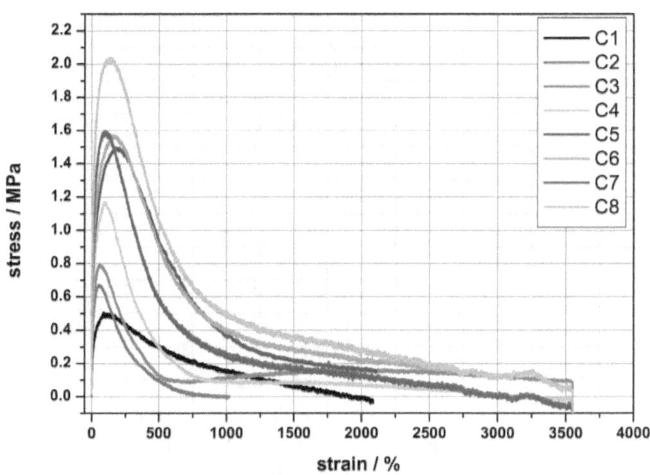

Figure 5.7.: Stress–strain-curves for nanocomposites of matrix polymer and
MMT modified with polymers of different polymerization degree.
All samples contain 5 wt.-% MMT.

by a broad region of plastic flow. The general shape and course
of the curves is the same. The height of the peak increases with
increasing grafted polymer chain length. By taking into account
how the addition of higher amounts of grafted MMT changes the
course and shape of the curves (see section 5.2) and comparing the
effect that had on the mechanical properties to this case where the
overall MMT content is kept constant it can be concluded that the
filler content is mainly responsible for the ability of the polymer
chains to flow. The increase in tensile strength and *Young's* modulus
could be attributed to the hydrodynamic effect. It was shown that
despite the tensile strength and *Young's* modulus being increased
the toughness stays rather constant (section 5.2). When now varying
the grafted polymer chain length, the plastic flow is still possible
clearly observable by the behaviour of the sample after the initial

Table 5.5.: Values for *Young's* modulus E, tensile strength σ_y, stain at break ε_B, and toughness U_T of nanocomposites with constant MMT-PMA content but varying grafted polymer chain length. Representative stress–strain-curves are shown in Figure 5.7. The maximum elongation of the used machine is 3495% for the used sample length and if reached, is referred to by the labelling "max.".

Sample	grafted $\overline{M_n}$ $/\ 10^3$ g·mol^{-1}	E $/$ N mm^{-2}	σ_y $/$ N mm^{-2}	ε_B $/\ 10^3$ %	U_T $/\ 10^6$ J m^{-3}
C1	4.1	5.1 ± 1.1	0.5 ± 0.1	2.09 ± 0.12	3.7 ± 0.5
C2	12.1	26.1 ± 2.6	0.7 ± 0.1	0.92 ± 0.13	1.9 ± 0.1
C3	18.1	26.8 ± 1.6	0.8 ± 0.1	3.45 ± 0.12	4.0 ± 2.1
C4	22.0	39.4 ± 1.8	1.0 ± 0.2	max.	4.9 ± 0.9
C5	28.0	58.6 ± 0.9	1.6 ± 0.2	max.	12.1 ± 0.4
C6	32.3	66.0 ± 1.7	1.5 ± 0.1	max.	13.7 ± 0.2
C7	38.6	87.7 ± 2.3	1.7 ± 0.2	max.	13.3 ± 5.0
C8	43.0	106.1 ± 2.7	2.0 ± 0.1	max.	20.3 ± 3.4

peak. Nevertheless, a strong increase in tensile strength and *Young's* modulus is observed with increasing chain length of the grafted polymer. In contrast to the experiments with increasing filler loading this may not be attributed to an increase of volume having a reduced chain mobility around an increasing amount of particles but more likely to an enhanced entanglement of the matrix chains with the surface-bound chains which clearly is more pronounced at higher chain lengths. The entanglements are like physical cross-links and energy is needed to break them.[57] To achieve this a high amount of conformational changes is needed which results in a mechanically stable network formed through entanglements. The energy necessary to break it is proportional to the number of entanglements,[137] thus, proportional to the chain length. Therefore, more energy is needed for elastic deformation with increasing chain length of the grafted polymer. Once the network is broken, the matrix chains are free to move and a more or less uniform plastic behavior can be observed.

From DSC, a glass transition temperature for the used PMA of 10.8 °C could be obtained. The composites show a slightly higher T_g of about (12.8±0.5) °C and do not show a dependency on the surface grafted polymers chain length.

5.4. Influence of mechanical stress on the microscopic structure of the composites

In order to analyze the influence of the strain on the microscopic structure of the composites, a suitable experiment was set up: a sample was subjected to 10% strain and was then placed under an optical microscope. This process was repeated until the sample broke or no further change was observed. Figure 5.8 shows a series of micrographs with increasing strain. It can be seen that at about 120 % strain the particles seem to be breaking up and are torn apart in the direction of strain. After being ruptured all torn particles are oriented with one corner in the direction of strain and between the two remaining pieces a cavity seems to be formed. This area increases with increasing strain. As shown from the results in section 4.3, the MMT is mostly not present in monolayers but in irregular stacks of very few layers. What might happen when the external stress is applied is that two of these layers slip against each other and are eventually separated from each other resulting in two new particles. The cohesion between these layers seems to be therefore weaker than the anchoring of the polymer onto the surfaces as in the opposite case no rupturing of the particles but only a cavity around the particles caused by a separation of particles and polymer would be observable. The first conclusion that can therefore be drawn from this experiment is that the anchoring of the polymer *via* VB16 to the surface is quite

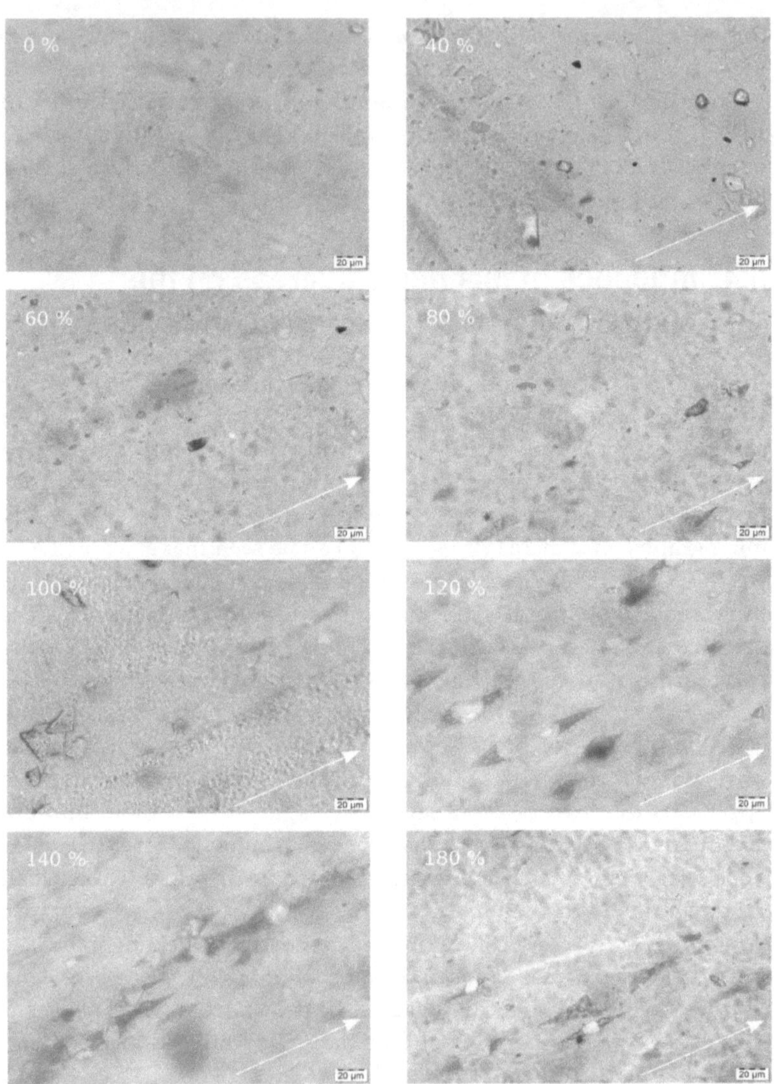

Figure 5.8.: Stepwise straining of a MMT-PMA filled PMA sample. Given percentages refer to the strain in respect to initial length and arrows indicate the direction of strain.

strong and can withstand external forces and that the first mechanism
leading to internal fractures is the shearing of particles.

This supports another observation that was often made throughout
the experimental part of this book: at strains higher than 100 %,
strain whitening was observed (Figure 5.9). Strain whitening is
an effect that results in the sample transforming from clear and
transparent to a white, milky, non-transparent color. This is often
explained with the formation of micro fractures in the sample.[138–140]
Within this experiment, the origin of these micro fractures could be
found. Both cases were examined further to show the respective
effect on the mechanical properties.

What remains open so far is the question how this behaviour of
the particles when subjected to an external force is influencing the
mechanical properties. Based on the theoretical background if the
number of layers in each stack is reduced this should only favor
the reinforcement of the material as the aspect ratio increases with
decreasing thickness of the particles. If, as described before, it can be
assumed that the anchoring of the surface grafted chains is still intact,
then no negative effect on the mechanical properties is expected,
when the material is recycled and used again to cast new specimen.
A different situation might occur when a specimen is strained over

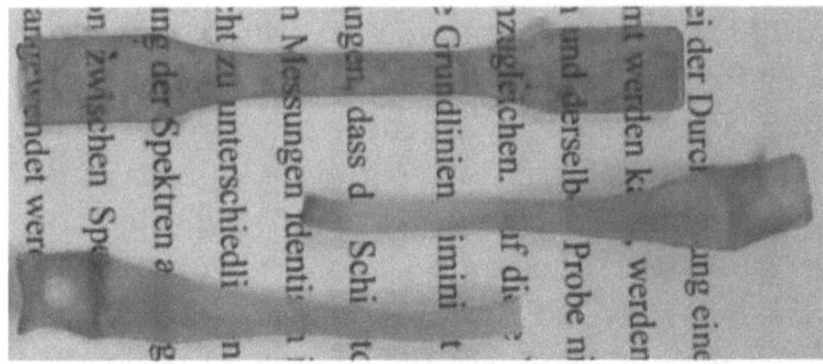

Figure 5.9.: Strain whitening effect on PMA filled with PMA-MMT: sample
before (top) and after (bottom two) tensile testing.

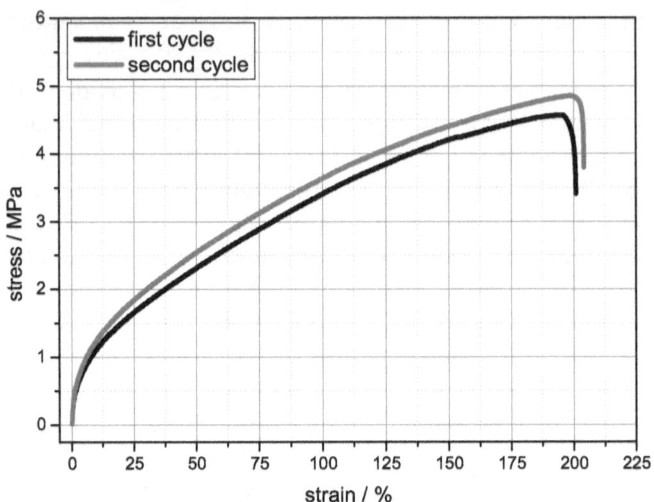

Figure 5.10.: Stress–strain-curves of two nanocomposites made from the exact
same material: first and second cycle.

a critical point where most particles are ruptured but the specimen
did not break and is then allowed to relax and tested again. In this
case, the aforementioned micro fractures may have an effect on the
mechanical performance.

Figure 5.10 shows two stress–strain-curves. The black curve is the
one recorded for the very first tensile testing that was performed with
this material. The blue one is the curve recorded for a sample made
from the same material by dissolving the composite in PGMEA and
re-casting of a testing specimen. Within the expected error margins
for tensile testing both curves are identical since both show the same
course and have nearly identical characteristic values. It can therefore
be concluded that, as expected, the change in particle morphology
has no negative influence on the materials properties.

Figure 5.11 shows three stress–strain-curves each for the nanocom-

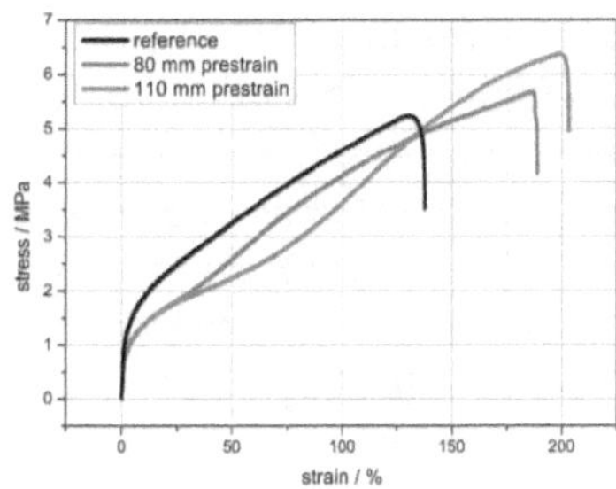 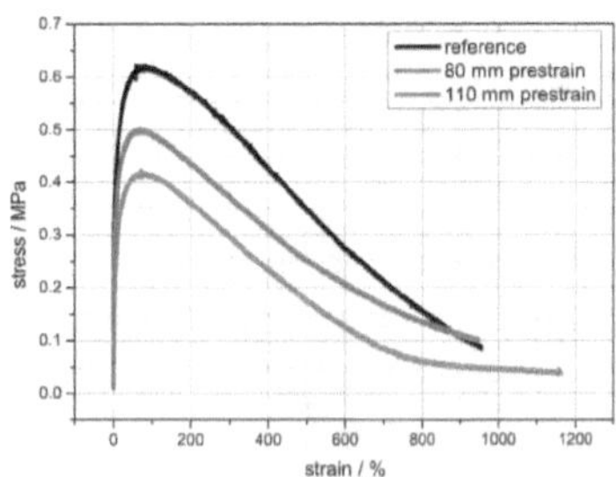

Figure 5.11.: Stress–strain-curves of composites (left) and pure PMA (right). Each figure shows a reference curve of a conventionally tested specimen and two curves of samples that were prestrained to 80 mm and 110 mm, respectively, prior to testing until failure.

posite (left) and for pure PMA (right): again one reference sample of the nanocomposite that was tested conventionally. The two other curves represent samples that were pre-strained to either 80 mm (60% strain) or 110 mm (120% strain), were put on a planar surface for one hour to rest and recover. Subsequently, they were then tested again until failure. 80 mm represents hereby a strain below the strain where all particles are deformed, while 110 mm is the strain found to be the critical threshold where nearly all visible particles are ruptured. For the PMA without particles the same experiment was performed in order to compare both. The stress–strain-curves for the PMAshow minor differences in the mechanical performance depending on the pre-strain, while the curves of the nanosheet filled composites clearly show a different behaviour. For both sets of samples it can be seen that the absolute stress at a given strain is slightly lower the higher the pre-strain was. However, it has to be noted that both graphs are scaled differently and the differences observed for the composite are significantly more pronounced. The fact that this effect occurs for both the PMA and the composite indicates that this might be attributed to a change within the polymer matrix. One explanation may be that after the sample is elongated for the first time, the chains are already partly orientated and disentangled. If this orientation is not completely reversed during the relaxation period, a lower stress at the same - or any given strain - would be expected as there is less energy needed to disentangle the chains. This is further supported, when analyzing the shape and course of the stress–strain-curves. The three curves of the pure PMA are all identical in shape and course. However, this is not the case for the three curves of the composites. Here, significant differences are visible. While the sample that was not pre-strained shows a constant rate of increase of the stress with increasing strain, both curves of pre-strained samples show a distinct flowing of the polymer. This is characterized by a deviation of the stress–strain-curve to lower stresses in the beginning of the plastic region. In this region, less force is needed to achieve the same strain. This effect is more pronounced, the higher the pre-strain is. The fact that only the course of the curves of the samples that contain particles changes may indicate that the origin of this effect can be found in the change of the particles due to the pre-strain. However, at a pre-strain of 80 mm the particles were not ruptured yet, so this is probably not the reason for the observed changes. One theory that could explain the experimental results, may be that due to the pre-strain, some chains are - partly or fully - disentangled. Despite the fact that the material was allowed to rest and took its original shape and size and that the polymer is above its glass transition temperature, maybe this disentanglement is not fully reversed. Therefore, less strain is needed initially for plastic deformation. As soon as the strain corresponding to the initial pre-strain is reached, the curves mirror the shape and course of the curves of the non-pre-strained samples. Nonetheless, the change in course of the curve is not observed for the samples without particles. This may hint that due to the presence of the particles, the chain mobility at room temperature is reduced and the chains are not able to return to their initial state after pre-straining which may be why this effect is only observed

for the samples containing particles. It is remarkable that, despite
the denoted changes, the ultimately reached stresses are identical or
even higher than the ones for the sample that was not subjected to
any pre-strain. This effect is even more pronounced for the sample
with ruptured particles (110 mm pre-strain), which might be due to
the increased number of particles due to the pre-strain. Even though
less energy is needed to strain the composites after pre-strain, the
reinforcing character of the particles is strong enough to counteract
this effect.

5.5. Conclusion

Within this chapter the preparation of different kinds of PMA–MMT-
nanocomposite and the effect that the surface modified particles
have on the mechanical properties and the materials performance
have been presented. The use of particles to enhance the composites
performance has been evaluated regarding three different aspects:
firstly, it could be shown that the surface modification with polymer
has significant advantages over the modification with small organic
molecules or even no modification at all. The *Young's* modulus of a
composite containing 5 wt.-% of PMA modified MMT was shown to
be nearly three times as high as the modulus of the pure polymer. At
the same time, the tensile strength increased by a factor of 10. The
ductility of the material was drastically reduced, while the toughness was increased by a factor of three. By comparing the shape
and course of the stress–strain-curve of composites containing PMA
modified particles to curves of composites with unmodified particles,
it becomes obvious that the reinforcing mechanism is stronger when
polymer modified particles are used. This is most likely attributed to
the fact that the surface bound chains entangle with the matrix, as
they are chemically identical, and to therefore prevent plastic flow
at much larger length scales. As it was shown by this experiment,

the concept of using polymer modified particles is a valid strategy to modify and tune the mechanical properties of composite materials and it was therefore examined further. In a next step, the dependency of the mechanical performance on the filler content was evaluated. Therefore different nanocomposites with a particle content varying up to 23 wt.-% were prepared. It could be shown that the variation of the filler content is a powerful tool to control the mechanical properties of the composite. Already small amounts of fillers result in a significant increase in *Young's* modulus and tensile strength. The addition of 22.6 wt.-% resulted in an increase in *Young's* modulus from (11 ± 1) N mm^{-2} to (5.2 ± 1.4) 10^2 N mm^{-2}. At the same time, the plastic properties of the material change completely. The strain-at-break is reduced to about 3% of its original value. The material is therefore transformed from a very ductile, soft material to a more rigid, stiffer material. The toughness remained rather constant, as increase in tensile strength and decrease in strain-at-break balance each other out. The reinforcing mechanism could be attributed to a reduced chain mobility which could occur due to the formation of a glassy layer around the particles. With increasing particle concentration, the proportion of chains with reduced mobility is increased and the overall stiffness of the material is increased. This was supported by DSC and DMA measurements. DSC showed a second glass transition that may originate from the polymer with reduced mobility. DMA showed a decrease in loss factor and therefore reduced viscous properties of the composite with increasing filler content. Creep measurements additionally showed the time dependent behaviour. All samples show time dependent increase of the strain which is in good agreement with the results from tensile testing and DMA.

Furthermore, the chain length of the grafted polymer was varied between $4.1 \cdot 10^3$ g mol^{-1} and $43 \cdot 10^3$ g mol^{-1}. *Young's* modulus and tensile strength showed a nearly linear growth with increasing grafted chain length while the matrix' degree of polymerization was kept constant. Both could be increased by one order of magnitude. The

plastic deformation and therefore the ability of the polymer chains to
flow freely, was not hindered, as all samples showed high strain-at-
break values. The toughness could in consequence be increased by
one order of magnitude over the range of the studied samples. As
plastic flow over a wide range is still possible, it could be concluded
that the grafted chain length variation has a localized effect around
the particles and the enhance of the mechanical properties originates
from improved interaction between particles and matrix.

Additionally, the effect of an external strain on the particles in the
composite has been evaluated by combination of optical microscopy
and tensile testing. Samples containing PMA modified MMT were
step wise strained and the particles were optically analyzed. Starting above 100 % rupturing of the particles in strain direction was
observed. It could be shown that samples with pre-ruptured particles
showed less resistance against plastic deformation at given strains.
Lastly, it could be shown that recycling of the samples to new testing
specimen did not result in samples with inferior mechanical properties. This could be attributed to the fact that the particles themselves
are still intact, only exfoliated. From theoretical approaches it is
known that an increase in aspect ratio is favorable for the improve-
ment of the mechanical properties. The observed effects do therefore
not have a negative impact.

Chapter 6

Matrix-free Montmorillonite-Nanocomposites: Synthesis and Properties*

Contents

Conventional Polymer–Nanocomposites, as for example those presented in chapter 5, consist of a nano-sized inorganic filler and a

---

*Parts of this chapter are part of the bachelor book "Synthese von Polymer–Montmorillonite-Filmen: Strategien zur Modifikation der mechanischen Eigenschaften" by Jytte Möckelmann (Georg-August-Univeristät Göttingen, **2019**).

polymeric matrix. Polymer-modification of the filler can be performed in order to improve the interaction between filler and matrix. Increasing mechanical properties are expected at an increasing filler content. When the matrix polymer content is reduced to zero, "matrix-free" composites, consisting only of particles and surface bound organic molecules, can be obtained. *Tchoul et al.* prepared titanium dioxide–polystyrene matrix-free composites by combination of phosphonate coupling and "click" chemistry.[141] They obtained composites with inorganic contents up to 80 wt.-%. *Choi et al.* synthesized both poly(styrene)- and poly(methyl methacrylate)-grafted silica nanoparticles and produced particle brush solids which are also matrix-free hybrid systems. They could observe a dependency of the *Young's* modulus, the fracture toughness, and the hardness on the degree of polymerization.[142] Crucial for the successful preparation of such a material is the presence of interactions between different particles trough entanglement of polymer chains bound to different particles. This is given, if the chains are long enough to entangle with each other or if the chains are cross-linked.[143] Additionally, research strives for achieving an order of the particles as it is hoped to produce materials with for example superior optical properties.[142,143]

Matrix-free MMT–polymer systems are prepared by imitating the biological structure of nacre. Nacre consists of 95 vol.-% aragonite platelets, which are a modification of calcium carbonate, and 5 vol.-% organic biopolymer.[13] They form a composite with a "brick-and-mortar" like structure which exhibits exceptional mechanical properties. The *Young's* modulus is known to be up to 135 GPa accompanied by a fracture toughness of up to $1.8 \, \mathrm{MJ \, m^{-3}}$. These properties can be attributed to the unique structure of the material.[13] A biomimetic approach to produce a comparable material is the use of layered nanoparticles to replace the aragonite and synthetic organic molecules to replace the biopolymer. Different nanoparticles, e.g. graphene oxide,[144,145] Al_2O_3 platelets,[146] or nanoclays[36,69,97,147] have been used so far.

In this chapter, the successful production of nacre-like MMT–polymer films will be presented and different strategies to tailor the mechanical properties by modification of the polymer functionalization of the platelets will be explored.

6.1. Synthesis of biomimetic nacre-like composites

A literature known procedure to produce thin MMT films is the evaporation of a dispersion of functionalized MMT from solution in a mold. In order to find a suitable solvent, PMA modified MMT was synthesized following the strategy presented in section 4.3. The free polymer was separated from the polymer modified nanosheets *via* centrifugation. The particles were then redispersed in different solvents. The dispersions were transferred into teflon molds and evaporation of the solvent was allowed over night. Figure 6.1 shows the results for films made from DCM, acetone, dimethoxyethane (DME), and PGMEA. It can be seen that the lower the vapor pressure of the chosen solvent the better are the results (Figure 6.1). PGMEA was therefore chosen as solvent for the film preparation and the evaporation was slowed down further by covering the samples. After 48 h the molds were placed into a vacuum oven at 100 °C and over the course of 8 h the pressure was reduced to <50 mbar, followed by drying *in vacuo* for 15 h to yield continuous, crack-free films (Figure 6.2).

Prior to the mechanical testing, all films were cut into approximately 5 mm wide stripes. All results that are presented in this chapter are obtained from films prepared as described.

Figure 6.1.: Films made from DCM, Aceton, DME and PGMEA.

Figure 6.2.: Continuous, crack-free thin film of PMA modified MMT from PGMEA.

6.2. Analysis of thin films with scanning electron microscopy

Figure 6.4 shows scanning electron microscopy (SEM) micrographs. During a SEM analysis electrons are accelerated towards the sample. While most of them are scattered, some are absorbed, which results in electrons of the sample being emmitted. They carry information about e.g. the samples topology or composition and can be analyzed. This process results in a charging of the sample. If the sample itself is not conductive, the charges remain roughly at the point where they reach the surface. This leads to overcharge effects and prevents analysis of the sample. The prepared MMT thin films are highly insulating and therefore need to be coated with a conductive layer,

Figure 6.3.: Sample preparation for SEM.

in this case gold. As the cross section of the samples was of highest interest, a film was cut using a scalpel and then wrapped around a sample holder (Figure 6.3). The sample stage of the SEM can be tilted for a few degrees so that it was possible to analyze both the cross section and the surface.

Via SEM it is not possible to visualize individual MMT layers, which are only about 1 nm in thickness, but only stacks of few MMT sheets. It can be seen that the cross section and the surface show structural differences (Figure 6.4, top). The surface of the sample seems to be relatively smooth, while the cross section is rough. This supports the idea that the MMT layers arrange parallel to each other. The cross section of the sample therefore represents also the cross sections of the MMT layers. Figure 6.4 (bottom) gives an additional impression of the layer-like structure of the cross section. The parallel aligned layers can be seen.

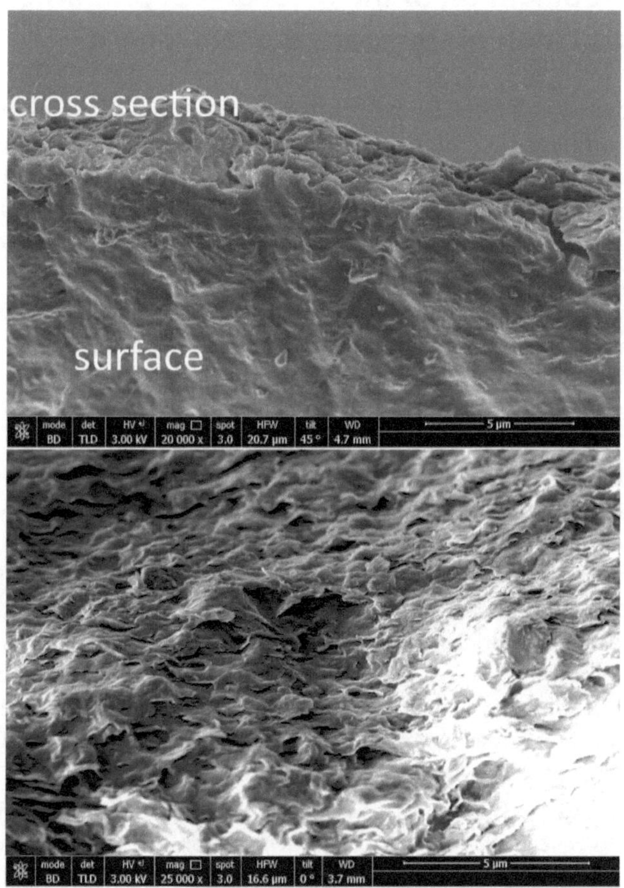

Figure 6.4.: SEM micrographs of an MMT film showing the surface and the cross section (top) and the layered structure of MMT nanosheets aligned parallel on the surface.

6.3. Modification of the mechanical properties *via* variation of the grafted chain length

As shown in chapter 5, the grafted polymer's chain length has a strong influence on the mechanical properties of the resulting composites. In a first approach, MMT was therefore modified with PMA of different molecular weight as presented in section 4.3 in order to determine its influence on the mechanical properties of the resulting films.

The SEC curves are shown in Figure 6.5 and the number average molecular masses and respective dispersities are presented in Table 6.1. It can be seen that the variation of the polymerization time resulted in a set of samples with a varying grafted polymer chain length between $0.4 \cdot 10^4 \, \mathrm{g \, mol^{-1}}$ and $4.3 \cdot 10^4 \, \mathrm{g \, mol^{-1}}$. The dispersities allow the conclusion that all polymerizations were carried out in a controlled fashion. Additionally, TGA was performed to analyze the polymer and MMT content of each sample. The curves are shown in Figure 6.6 and the calculated polymer content is given in Table 6.2. It can be seen that with increasing grafted polymer chain length the polymer content increases as well. With constant VB16 content the same number of grafted chains is expected. Variation of the polymerization time or increase in molecular mass does therefore result in an increase in organic content. This has to be kept in mind when interpreting the results of the mechanical testing.

Additionally, TGA analysis reveals that the polymer modification seems to result in an enhanced thermal stability of the composite. The step at about $250\,^\circ\mathrm{C}$ can be attributed to the decomposition of the aliphatic part of the VB16 (see also subsection 4.1.1). This decomposition seems to be shifted to higher temperatures as a consequence of the coating with PMA.

Figure 6.7 shows representative stress–stain-curves. From sample **L-I**, no continuous film could be produced because the films were

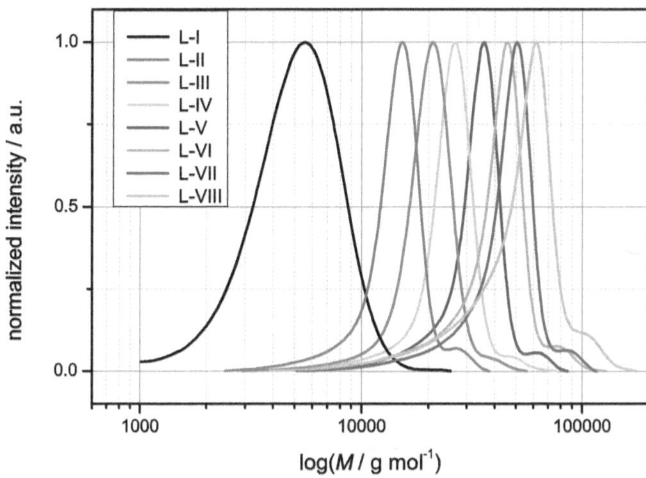

Figure 6.5.: Normalized molar mass distributions of the free polymer of polymerizations **L-I** to **L-VIII**. All curves are obtained by evaluation of the RI detector. The respective number averages and dispersities are given in Table 6.1.

too brittle. This gives rise to the assumption that a minimal grafted polymer chain length is required to form continuous films. In order to produce such films, the polymer chains on the particles need to be able to interact not only with themselves or chains grafted onto the same particle but also with chains grafted onto other particles. For polymers, the entanglement molecular weight describes the limit above which chains are able to entangle with each other, while below that limit the chains are described as rods. For PMA, the entanglement molecular weight is known to be approximately $1.1 \cdot 10^4 \, \mathrm{g \, mol^{-1}}$.[136] As the molecular weight of **L-I** is below that limit, no entanglement between chains is expected which is confirmed by the fact that no continuous film could be produced.

In Figure 6.7 it can be seen that the variation of the grafted poly-

Table 6.1.: Number average molecular masses and dispersities of the free
polymer of the polymerizations for preparation of films **L-I** to
L-VIII. The respective distributions are shown in Figure 6.5

sample	$\overline{M_\mathrm{n}}$ / $10^4 \mathrm{g\,mol}^{-1}$	$Đ$
L-I	0.4	1.3
L-II	1.2	1.2
L-III	1.8	1.1
L-IV	2.2	1.2
L-V	2.8	1.2
L-VI	3.2	1.3
L-VII	3.9	1.2
L-VIII	4.3	1.3

mer's chain length has a strong influence on the mechanical properties of the films. First of all, the tensile tests show that the mechanical
properties of the films are strongly determined by the MMT as the

Table 6.2.: TGA results of films **L-II** to **L-VIII**. Additionally, the grafted chain
length is given again for comparison. As from sample **L-I** no
continuous film could be produced, it is not considered for further
analysis.

sample	$\overline{M_\mathrm{n}}$ / $10^4 \mathrm{g\,mol}^{-1}$	Polymer content / wt.-%
L-II	1.2	52.73
L-III	1.8	53.34
L-IV	2.2	57.23
L-V	2.8	64.41
L-VI	3.2	64.99
L-VII	3.9	67.37
L-VIII	4.3	71.38

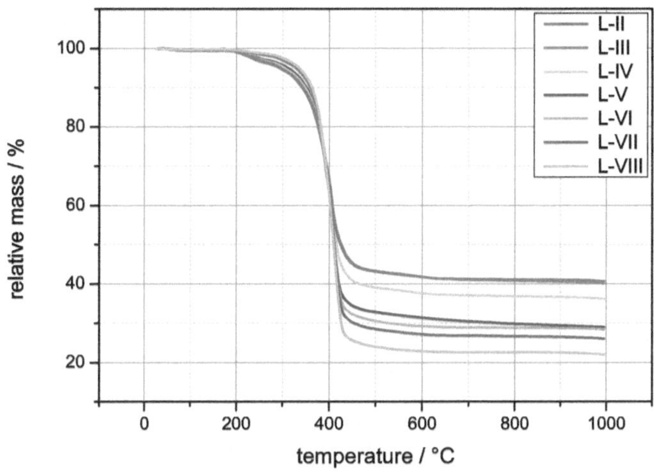

Figure 6.6.: Thermogram of samples **L-II** to **L-VIII**. As from sample **L-I** no
continuous film could be produced, it is not considered for further
analysis.

Young's modulus of the analyzed films is between 50 and 550 times
higher than the *Young's* modulus of pure PMA. None of the curves
shows pronounced plastic flow of the polymer chains, which would
be observable as a maximum in the stress strain curve followed by a
decrease in stress with increasing strain as it is strongly pronounced
for PMA.

From the stress–strain-curves it can be determined that the behaviour of the films changes depending on the grafted polymer's
chain length. The characteristic values for the *Young's* modulus, the
tensile strength, strain at break, and toughness depending on the
grafted polymer's chain length that are obtained from tensile testing,
are shown in Table 6.3. All data given are averages of eight to twelve
measurements. Errors are calculated as standard deviation. It can be
seen that the *Young's* modulus of the films decreases with increasing

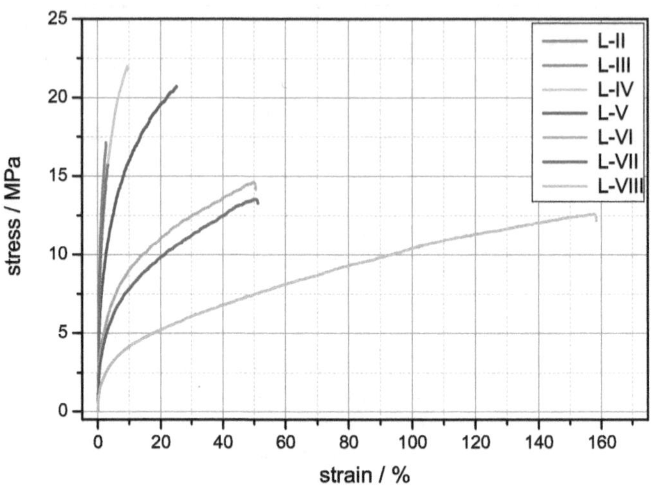

Figure 6.7.: Representative stress–strain-curves of **L-I** to **L-VIII**.

grafted molecular weight. At the same time both the toughness
and the strain-at-break increase with increasing grafted polymer's
chain length. This behaviour can be explained as follows: Each film
consists of the same amount of MMT. So if the grafted chain length
is increased, the overall polymer content increases. The MMT fraction in the film is therefore decreased. The influence of the softer
PMA increases, consequently resulting in a decrease in the *Young's*
modulus. At the same time, as the polymer chains get longer the
interactions between them increase. When the film is exerted to an
external uniaxial tensile force, the interacting chains withstand this
force and the material starts to flow instead of breaking. Hence the
material can withstand increasing strains by deforming. The higher
the polymerization degree, the more pronounced does this plastic
deformation get. It can therefore be concluded that the MMT is re-

Table 6.3.: *Young's* modulus E, strain at break ε_B, tensile strength σ_y, and toughness U_T of MMT-PMA films with variable grafted polymer's chain length.

sample	E / 10^2 N mm^{-2}	ε_B /%	σ_y /MPa	U_T / 10^6 J m^{-3}
L-II	22.1 $\pm$ 2.3	3 $\pm$ 1	18.0 $\pm$ 2.1	0.4 $\pm$ 0.1
L-III	21.5 $\pm$ 1.2	3 $\pm$ 1	17.7 $\pm$ 1.4	0.4 $\pm$ 0.1
L-IV	18.5 $\pm$ 1.5	9 $\pm$ 1	22.8 $\pm$ 1.1	1.5 $\pm$ 0.2
L-V	12.5 $\pm$ 1.3	26 $\pm$ 4	20.7 $\pm$ 1.5	4.2 $\pm$ 0.8
L-VI	6.69 $\pm$ 0.17	50 $\pm$ 10	14.6 $\pm$ 1.3	5.7 $\pm$ 1.5
L-VII	5.56 $\pm$ 0.32	68 $\pm$ 14	14.8 $\pm$ 1.5	7.6 $\pm$ 2.1
L-VIII	2.76 $\pm$ 0.12	154 $\pm$ 26	12.5 $\pm$ 1.1	13.7 $\pm$ 3.2

sponsible for the materials *Young's* modulus and the grafted polymer for the plastic deformation. In contrast to thermoplastics above their glass transition, as for example PMA, where the chains can move freely when an external stress is applied, all samples show a distinct breaking point. In all cases, the highest stress that could be applied to the sample was also the stress-at-break. As a combination of this and the increasing strain-at-break, the toughness of the material increases with increasing grafted chain length or polymer content.

Eventually, in the used system it is not possible to separate the effect of the grafted chain length and the MMT content, as one causes the other, but there are indications that give rise to the assumption that the observed effects are not only caused by variation of the MMT content. Due to the fact that a *grafting through* approach is applied, it is not possible to precisely control the number of polymer chains that are bound to the surface during the polymerization. This causes samples **L-V** and **L-VI** to have approximately the same MMT content at different average molar masses. Their mechanical properties, however, are significantly different. All evaluated characteristics follow the trend observed for all samples regardless of their same MMT content. This indicates that the MMT-content is not solely causing the observed effect but rather a combination of both influences. Nevertheless, changing the molecular weight of the grafted polymer is a very easy strategy to precisely tune the mechanical properties of the resulting composite.

6.4. Mixing approach for the formation of films

The results from section 6.3 showed that the highest *Young's* modulus would be expected for films made from particles functionalized with polymer of a very low molecular mass or, in other words, for films with a very high MMT content. It could be shown, however, that this kind of particles are not able to form continuous films because of the lack of interaction between the polymer chains. A strategy to overcome this limitation was developed using a mixing approach of two different kinds of particles. Mainly nanosheets functionalized with polymer of a chain length too short for film formation were used (sample i) and particles modified with polymer of a very high molecular mass which are known to work well in film formation were added in small amounts (sample ii). The average molar masses and dispersities are given in Table 6.4. Table 6.5 gives the proportions of particles modified with short chains and long chains respectively, for all samples investigated.

The films were then characterized with TGA to evaluate the MMT content which has, as shown in section 6.3, a huge impact on the mechanical properties. Generally, with increasing proportion of parti-

Table 6.4.: Average molar masses for the films produced by a mixing approach.

sample	$\overline{M_n}$ / 10^4 g mol^{-1}	$Đ$
i	0.87	1.3
ii	2.98	1.1

Table 6.5.: Proportions of particles modified with short chains and long chains, respectively, in the films produced *via* the mixing approach. The proportions are calculated on percentage of MMT.

sample	short chains / wt.-%	long chains / wt.-%
mix-0	0	100
mix-60	60	40
mix-70	70	30
mix-80	80	20
mix-90	90	10

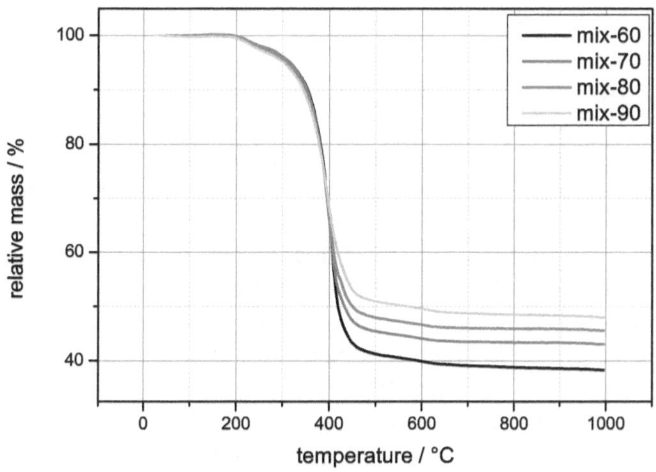

Figure 6.8.: Thermogram of different films produced *via* the mixing approach.

cles with short grafted polymer chains an increasing MMT content is observed, as has been expected. Due to the application of this mixing approach, the inorganic content could be increased to about 55wt.-%.

In order to evaluate the mechanical properties, tensile testing was performed. From the stress–strain-curves (Figure 6.9) it can

Table 6.6.: TGA Results for films **mix-60** to **mix-90**.

sample	fraction of short chains / %	organic content / wt%
mix-60	60	55.15
mix-70	70	54.44
mix-80	80	47.84
mix-90	90	45.40

be concluded that the addition of particles with short grafted chains has a major impact on the mechanical performance of the material. Especially the maximum strain decreases significantly. Film **mix-90** shows nearly no plastic deformation at all. At the same time, it also exhibits the highest *Young's* modulus of any of the produced films. The exact determined parameters can be seen in Table 6.7. Here, the huge impact of the mixing of the two types of particles is clearly visible. While the *Young's* modulus increases nearly linearly with increasing amount of particles grafted with short chains, both the toughness and the strain-at-break decrease. Interestingly, the tensile strength does not show this behaviour. Over the studied range, the obtained values were independent of the content of short chain grafted particles. For a practical application, this means that by applying the presented mixing approach, a material with a high tensile strength and completely freely tunable *Young's* modulus can be produced. Up to here, these quantities were always connected. For the films with only one kind of anchored polymer (section 6.3) a change in *Young's* modulus always meant a direct change in tensile strength. Now it is possible to change the *Young's* modulus while maintaining the high tensile strength.

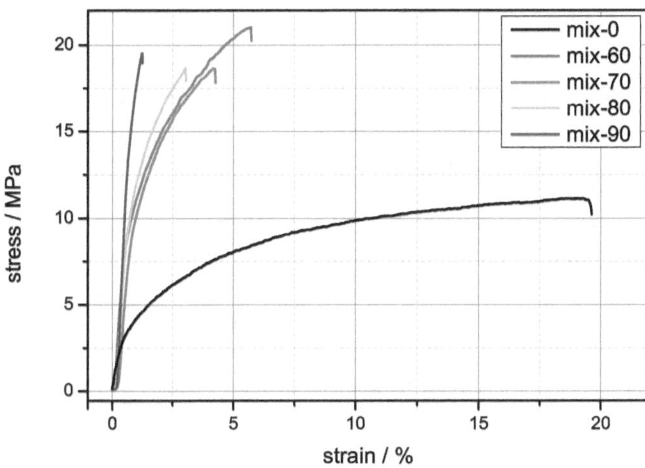

Figure 6.9.: Representative stress–strain-curves of films **mix-0** to **mix-90**.

Table 6.7.: *Young's* modulus E, strain at break ε_B, tensile strength σ_y, and toughness U_T for filmes prepared *via* mixing approach.

sample	E / 10^2 MPa	ε_B /%	σ_y /MPa	U_T / 10^6 J m^{-3}
mix-0	7.8 $\pm$ 1.0	18.9 $\pm$ 4.6	10.7 $\pm$ 0.8	1.6 $\pm$ 0.5
mix-60	19.2 $\pm$ 2.5	7.7 $\pm$ 2.3	19.7 $\pm$ 1.9	1.2 $\pm$ 0.4
mix-70	21.9 $\pm$ 1.6	3.9 $\pm$ 1.0	18.8 $\pm$ 2.4	0.55 $\pm$ 0.20
mix-80	25.4 $\pm$ 2.6	3.1 $\pm$ 0.8	19.7 $\pm$ 3.1	0.44 $\pm$ 0.20
mix-90	34.1 $\pm$ 2.7	1.3 $\pm$ 0.4	18.6 $\pm$ 2.9	0.16 $\pm$ 0.05

6.5. Modification of the mechanical properties *via* cross-linking of the grafted polymer

The introduction of cross-links is a well known strategy to tune the mechanical properties of a polymer. Chemical cross-linking can be performed both during and after a polymerization. The usage of tri-functional comonomers during a radical polymerization results in random cross-links throughout the polymer chain. One disadvantage of this strategy is that cross-linked polymers are usually not soluble but only swellable. Further processing, as for example the solvent casting that is performed in this book, is therefore not possible. Another strategy to introduce chemical cross-links is the addition of comonomers that can be cross-linked after the polymerization. One example is the thermo-reversible cross-link of polymers using Diels-Alder reactions.[148,149] Another strategy for the introduction of cross-links is a change in topology of the polymer by adding star polymers. While such systems are no cross-links in the traditional sense, it has been shown that star polymers have a higher effective bridge connection in copolymers.[140,150] It can therefore be concluded that the interaction between different chains is enhanced.

Previous works from this group have shown that the introduction of physical cross-links also results in a drastic change in mechanical properties.[127,151] It was demonstrated that the introduction of hydrogen bonding moieties to acrylate polymers results in a significant increase of the toughness and allows for a healing effect. This approach offers the advantage that physical cross-links, after being broken, may be reformed.

Within this book, films consisting of nanosheets coated either with polymer containing various amounts of star polymer and or polymer with hydrogen bonding sites are synthesized and analyzed to explore the effect of cross-linking on their mechanical properties.

6.5.1. Cross-linking *via* star-polymer

The chosen star RAFT agent (Figure 6.10) is a three-arm star RAFT agent.[152] Using it in combination with the former applied linear 2-(dodecylthiocarbonothioylthio)propionic acid (DSPA) results in linear polymer and star polymer. During the polymerization it is expected

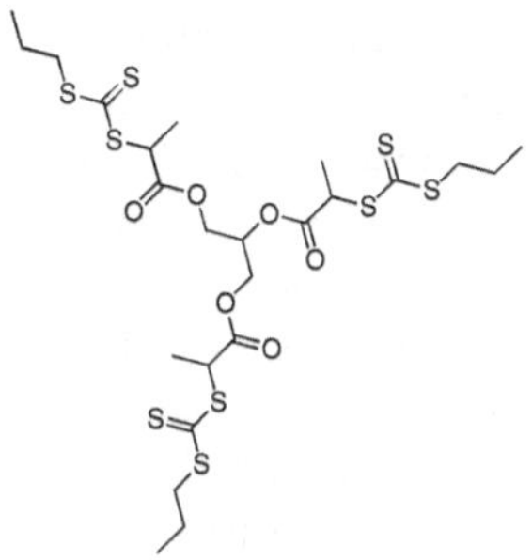

Figure 6.10.: Star RAFT agent S3.

that radicals have the same probability to add to a TTC group of a star or as to linear RAFT agent.[140] Consequently, each arm of the star polymer and the linear polymer are expected to have the same degree of polymerization. As both polymers are chemically identical but only differ in their topology, the star cores can be understood as cross-links of linear PMA. Cross-linking of linear polymer enhances the restoring forces within the network under external stress as the movement of chains is hindered and they are held in place. The ability of the material to deform under stress is therefore reduced. In consequence, a higher *Young's* modulus and a reduced strain at break are expected.

In order to investigate the influence of the star polymer on the mechanical properties, polymerizations of methyl acrylate (MA) using various proportions of star RAFT agent were carried out in the presence of MMT-VB16. The total concentration of RAFT groups within the systems were held constant for each polymerization and the fraction of RAFT groups that are part of a star RAFT agent was varied between 30 mol% and 100 mol%. Figure 6.11 shows molar mass distributions obtained *via* SEC. As can be seen, all of the samples show multimodal distributions. When analyzing the individual peaks of each distribution, it becomes apparent that they

are multiples of the first peak of each respective distribution. It is
known that due to transfer to polymer star-star-coupling is possible.
Boschmann and *Vana* proposed a mechanism for this that is based
on the assumption that in a first inter molecular step, the terminal
radical of a growing polymer chain is transferred one arm of a star,
resulting in a mid-chain radical. This radical is then able to continue
the propagation. As all growing radicals continuously undergo addition and fragmentation with any RAFT group, the growing mid
chain radical will eventually add to the RAFT core of another star,
resulting in a molecule consisting of two stars with hence twice the
molar mass.[153] Additional peaks in the distribution may originate
from further star-star-coupling. Additionally to the star polymer
there is also linear polymer formed. However, no peak can be easily
assigned to the linear polymer. Theoretically, the growing chains of
the linear polymer and the individual arms of the star polymer are
exchanged during the RAFT mechanism. It can therefore be assumed
that they are all of the same molecular mass. The first peak of each
distribution should therefore be originating from the linear polymer,
while all others originate from star polymer. This is further supported
by the observation that for the sample with the lowest star RAFT
content, this peak is the highest and decreases with increasing star
RAFT content. With increasing molar fraction of the star RAFT agent,
the number of RAFT groups in the system is increased and the ratio
of monomer-to-RAFT groups decreases in consequence. This can
be observed in the SEC curves as the first peak of each distribution
is shifted continuously to lower molecular masses as the star RAFT
content increases.

As the molar mass distributions are quite complex and it is impossible to determine one specific molar mass for each sample, for better
comparability of the samples not the average molecular weight but
rather the overall polymer loading of each sample is compared. TGA
yields a polymer content of 56.3 ± 1.2 wt.-% for all samples.

From all six samples films were prepared and tensile testing was

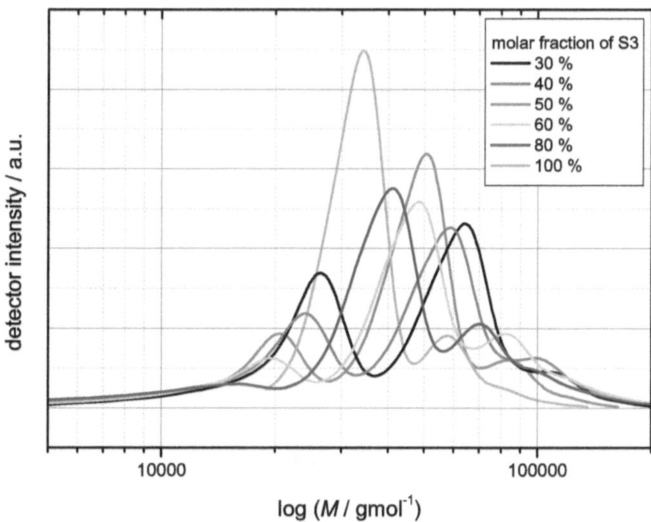

Figure 6.11.: SEC chromatograms of the free polymer of polymerizations for the functionalization of nanosheets using a mixture of linear and star RAFT.

performed. In Figure 6.12 the respective stress–strain-curves are shown. Table 6.8 gives the corresponding values for the average Young's modulus, tensile stress, strain at break, and toughness. Tensile testing clearly shows the impact of the change in polymer topology. Initially, when the star RAFT content is increased from 30 % to 60 % the *Young's* modulus of the samples is increased. This can be attributed to the increasing cross-link density within the polymer that enhances the material's ability to withstand external stress. This is also clearly visible in the results of the strain at break that is significantly reduced throughout this experiment, because the added cross-links prevent plastic deformation of the samples. The tensile strength of samples **S-I** to **S-IV** is rather constant. In consequence,

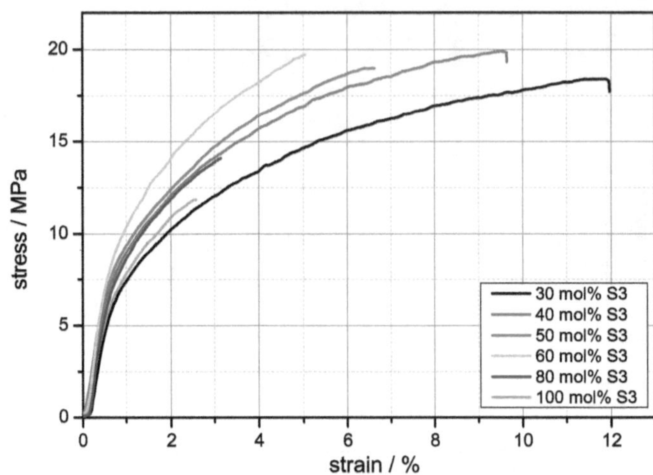

Figure 6.12.: Stress–strain-curves of films made from nanosheets coated with
varying combinations of linear and star polymer.

the toughness decreases, as would be expected for a more rigid material. When the star RAFT content is increased to 80 % or even 100 %, the *Young's* modulus and tensile strength decrease. A possible explanation might be that due to the high star RAFT content the average chain length is reduced, as discussed before. A consequence of this is that the chain's ability to entangle with each other is reduced. As discussed in chapter 5, this results in inferior mechanical properties.

Comparing these results to the results obtained for films made from nanosheets covered with linear polymer gives further insight as to how this strategy might be used to tune the mechanical properties of matrix-free MMT-PMA composites. Usually, a change in one property automatically results in a change of all the other properties. This can be observed for the samples with variation of the grafted polymer's chain length. In contrast, addition of star polymer allows for a change in some characteristics while keeping the tensile strength

Table 6.8.: *Young's* modulus E, strain at break ε_B, tensile strength σ_y, and toughness U_T of samples with varying star RAFT content.

sample	fraction S3 / mol %	E / 10^2 MPa	ε_B /%	σ_y /MPa	U_T / 10^6 J m^{-3}
S-I	30	15.7 $\pm$ 1.6	11.8 $\pm$ 2.2	20.5 $\pm$ 2.8	1.9 $\pm$ 0.4
S-II	40	16.4 $\pm$ 2.9	9.3 $\pm$ 1.9	20.0 $\pm$ 1.0	1.4 $\pm$ 0.4
S-III	50	16.3 $\pm$ 1.7	6.6 $\pm$ 1.9	18.6 $\pm$ 1.5	0.9 $\pm$ 0.4
S-IV	60	19.8 $\pm$ 2.1	4.7 $\pm$ 1.0	19.3 $\pm$ 2.2	0.6 $\pm$ 0.2
S-V	80	18.3 $\pm$ 2.3	2.4 $\pm$ 1.1	13.6 $\pm$ 2.5	0.2 $\pm$ 0.1
S-VI	100	14.79 $\pm$ 0.56	2.6 $\pm$ 1.9	11.2 $\pm$ 3.3	0.3 $\pm$ 0.3

rather constant, which is a very special property of these samples. It allows to create a material with one set property while the others can be freely adjusted. In order to further compare samples with star polymer and purely linear polymer, sample **L-IV** is suitable, as it has a comparable amount of grafted polymer. Modulus and tensile strength of the star-samples **S-I** to **S-IV** are relatively similar to this sample. It can be concluded that the main benefit of the addition of cross-links *via* star polymer is the control over the plastic deformation of the material that is now accessible.

6.5.2. Cross-linking *via* H-bonding comonomers

The introduction of hydrogenbonding moieties to polymers by using comonomers with H-bonding sites in their side-chains is a strategy that has already been shown to result in excellent mechanical properties.[151,154,155] It is known that these kind of polymers can absorb high amounts of energy when subjected to external stresses due to breaking and reforming of H-bonds. The outstanding advantage of this approach lies in the reversibility of this process: Once the external loading is ended, the flexible polymer chains relax and

Scheme 6.1: Chemical formula of 2-carboxyethylacrylate (CEA) that was used
as comonomer to introduce H-Bonds to the polymer modification
of MMT.

new H-bonds can be formed allowing for the same or even higher
mechanical resilience.[155]

This strategy has been adapted to tune the mechanical properties of matrix-free polymer–MMT-nanocomposites. In a first step,
H-bonding cross-linked MA-MMT composites are prepared and analyzed in order to add a further strategy to the beforehand presented
ones to tune the composites properties. Additionally, it is shown that
this approach opens up the access to matrix-free butyl acrylate (BA)-
MMT composites, which are otherwise inaccessible due to the low
glass transition temperature of BA.

6.5.2.1. MA-based composites

For MA-based matrix-free MMT composites 2-carboxythylacrylate
(CEA) (Scheme 6.1) was chosen as comonomer. As CEA is an acrylate,
the copolymerization with MA is expected to work well. Poly-CEA
is additionally known to have a low glass transition temperature
of $T_g \approx 22°C$ and its successful RAFT polymerization has already
been reported.[156] The fact that the glass transition temperature is
relatively close to the glass transition temperature of PMA allows
for the conclusion that all enhancing effects can be attributed to the
introduction of cross-links and not to the introduction of more rigid
parts in the polymer back bone.

In order to produce samples that can be compared to each other, it
is necessary to know the properties of the surface bound polymer. Up
to this point, throughout this book, the molar mass of the polymer

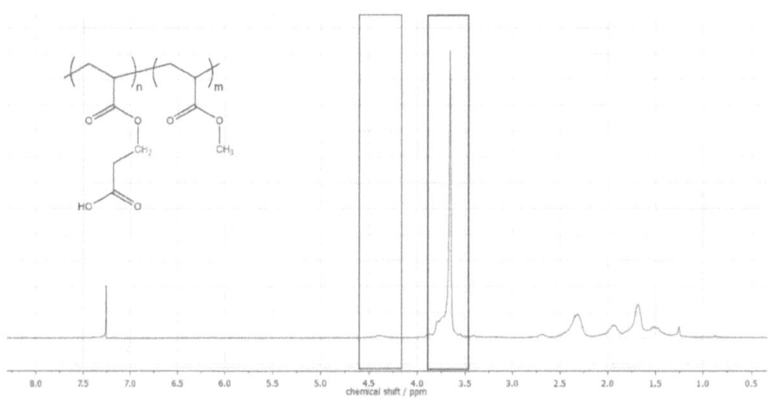

Figure 6.13.: NMR spectrum of a copolymer of MA and CEA. Red marks chosen peak for CEA, blue for MA.

was used to compare the polymer. However, for this polymer, it is not possible to conduct SEC experiments. Molar mass detection *via* SEC is based on the correlation between radius of gyration, molar mass and eluation time. The highly polar carboxyl-groups of the CEA result in interaction of the polymer with the columns of the SEC setup. Consequently, the eluation time of the molecules is increased, which prevents analysis on the available SEC setups. Fortunately, a combination of gravimetric analysis and nuclear magnetic resonance (NMR) spectroscopy allows for determination of an approximate molar mass and the CEA content. Both the free polymer and the polymer modified nanosheets were dried and weighed. The overall polymer content was calculated and the average molar mass was obtained. Figure 6.13 shows an NMR spectrum of a copolymer of MA and CEA. By comparing the integrals of the $-CH_3$ group of MA (approx. 3.6 ppm) and the $-CH_2$ group adjacent to the ester group of the CEA (approx. 4.4 ppm) and weighting with the respective amount of protons gives rise to the CEA content.

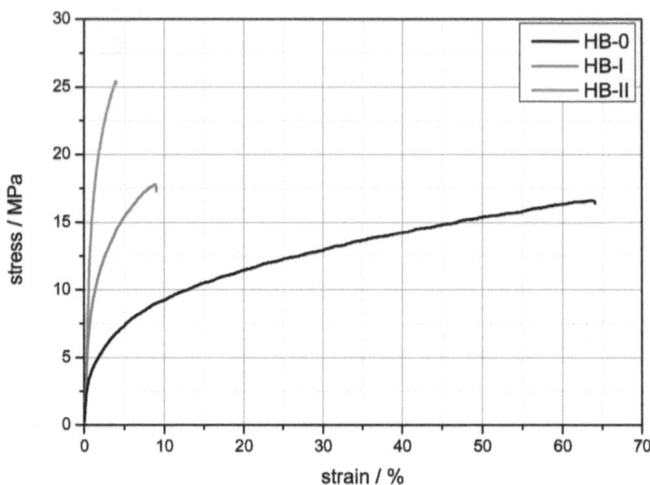

Figure 6.14.: Stress–strain-curves of CEA containing matrix-free MA-MMT
composites.

Figure 6.14 shows the stress–strain-curves of samples containing
3 mol% and 7 mol% CEA and as a reference, the respective curve
of a film made from nanosheets coated with pure PMA of the same
polymerization degree. The impact of the H-bonding moieties is
clearly visible: The *Young's* modulus could be increased by a factor
of four, while the strain at break was decreased by a factor of ten
and the tensile strength was doubled. The toughness decreases in
consequence. When put into relation to the results presented so
far, the best strategy to obtain a rigid material with a modulus this
high, was to reduce the polymer content as much as possible. The
introduction of H-bonds, however, opens up the possibility to keep
a higher polymer content, which may for example be associated
with better resistance against water, while maintaining extraordinary
mechanical properties.

Table 6.9.: *Young's* modulus E, strain at break ε_B, tensile strength σ_y, and toughness U_T of CEA containing matrix-free MA-MMT composites.

sample	CEA content mol%	E /10^2 MPa	ε_B /%	σ_y /MPa	U_T / 10^6 J m^{-3}
H-0	0	6.7 ± 0.2	50 ± 10	14 ± 1	5.7 ± 1.5
H-I	3	15.7 ± 1.1	10 ± 2	18 ± 1	1.3 ± 0.9
H-II	7	23.4 ± 5.4	5 ± 1	25 ± 4	0.7 ± 0.3

6.5.2.2. BA-based composites

During preliminary work to chapter 6, different polymers were studied to find a suitable polymer to produce self-supporting, matrix-free polymer-MMT composites. One of the polymers investigated was PBA. PBA has a literature-T_g of approximately $-54\,°C$, making it a very soft, ductile polymer at room temperature. While the modification of MMT nanosheets with PBA was successful, the produced films were still too soft to take them out of the molds they were made in. However, having already shown that H-bonding moieties result in the composite being more rigid and less ductile, this might be a promising strategy to also produce composite materials of MMT and polymers that were otherwise not accessible. In order to do so, BA and N-acryloyl-L-phenylalanine (APA) were copolymerized in the presence of MMT-VB16. APA (Figure 6.15) is a monomer that is a derivative of phenyl-L-alanine. It was chosen because it has one more hydrogen bonding site than CEA and is therefore expected to have a stronger reinforcement effect on the composite which may be beneficial to counteract the low T_g of PBA. Additionally, it has already been shown to have a comparable effect on the mechanical properties of polymers as CEA.[127] The preparation of films with varying APA content was successful and the films were mechanically stable enough to be taken out of the molds. The hydrogen-bonding comonomer in the grafted polymer results therefore in a reduction of

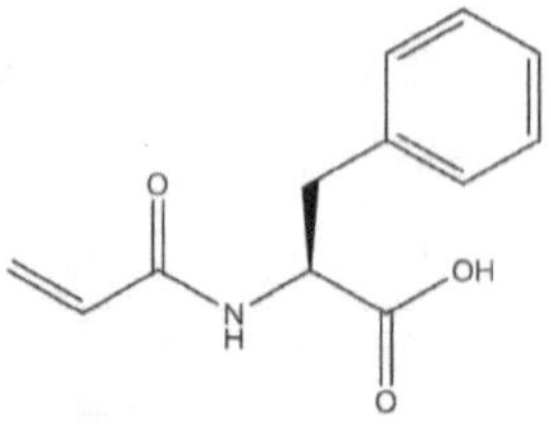

6.5. Modification of the mechanical properties *via* cross-linking of the grafted polymer

Figure 6.15.: Structural formula of APA.

Table 6.10.: *Young's* modulus E, strain at break ε_B, tensile strength σ_y, and toughness U_T of MMT-p(BA-*co*-APA) composites.

sample	APA content mol%	E / 10^2 MPa	ε_B /%	σ_y /MPa	U_T // 10^6 J m^{-3}
HB-III	1.1	0.31 ± 0.04	2.2 ± 0.8	2.6 ± 0.2	0.05 ± 0.02
HB-IV	2.8	4.6 ± 1.4	2.0 ± 0.4	4.4 ± 0.8	0.06 ± 0.01
HB-V	5.9	6.7 ± 2.0	1.6 ± 0.3	4.7 ± 1.6	0.06 ± 0.02
HB-VI	10.6	10.8 ± 1.9	2.1 ± 0.6	8.8 ± 2.0	0.12 ± 0.06

the chain mobility preventing plastic flow and allows for the formation of otherwise inaccessible "matrix free" MMT-PBA composites. Representative stress–strain-curves are shown in Figure 6.16 and corresponding data is given in Table 6.10. It can be seen that the effect of the APA on the composites properties is consistent with the results presented for H-bonding MA-MMT composites. This was expected, as this effect was solely attributed to the introduction of cross-links through hydrogen bonding. What can be highlighted here, however, are the remarkable properties of the composite made from MMT and a polymer approximately 80 K above its glass transition temperature. A peak Young's modulus of $(10.8\pm1.9) \cdot 10^2$ MPa and tensile strength of (8.8 ± 2.0) MPa are far above conventional PBA-nanocomposites and show the potential of this strategy.[157]

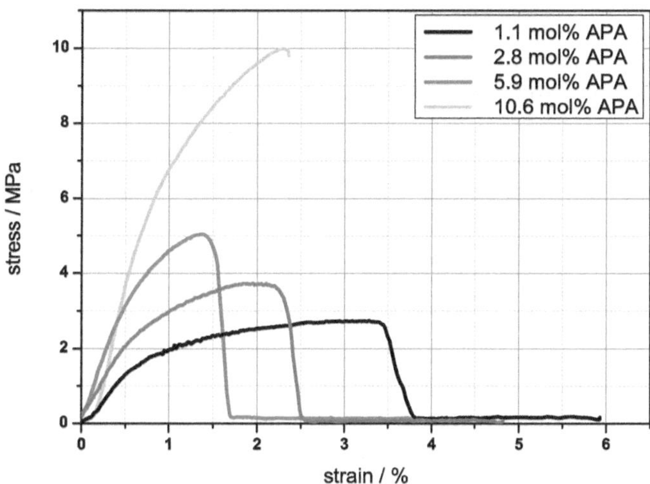

Figure 6.16.: Stress–strain-curves of MMT-p(BA-*co*-APA) composites with varying amounts of APA.

6.6. Conclusion

The successful synthesis of mechanically stable "matrix-free" films from polymer coated nanosheets was presented. The optimal conditions for the formation of crack-free samples were found and an impression of the surface and cross section structure could be gained *via* SEM imaging. Subsequently, different strategies to tune the mechanical properties of the films were developed and explored.

Based on the results of chapter 5, the first approach was to vary the grafted polymer's chain length. As the grafting density of the VB16 units was constant, variation of the grafted polymer's molar mass results in a variation of the polymer content of the composite particles. Films made from these particles therefore varied in polymer content or MMT content, respectively. This was confirmed *via* TGA.

TGA also revealed that an increasing polymer chain length results in more temperature stable particles. It could be shown that there is a minimal chain length necessary to form continuous, crack free films. This corresponds roughly to the literature known entanglement molecular weight of PMA. It was therefore concluded that in order to form films, chains need to be able to interact and entangle with each other. Subsequently, the mechanical properties of the films were investigated *via* tensile testing. The toughness of the produced materials was found to be in the same order of magnitude as of conventional composites with matrix and filler (see chapter 5), which is surprising given the extremely high filler content. This can be attributed to the extraordinarily high *Young's* modulus and tensile strength. Furthermore, it could be shown that in order to produce films with even higher MMT content, it is sufficient to have a small proportion of particles decorated with long polymer chains, while the majority of the nanosheets is covered with polymer of a low molecular weight. Additionally, cross-links were introduced to the grafted polymer using two different strategies: Cross-linking *via* a change in the polymer's topology through addition of star polymer and *via* the introduction of hydrogen-bonding sites. The introduction of various proportions of star polymer resulted in composites with even higher modulus. Remarkable was that with varying star polymer content on the surface of the particles that modulus stayed rather constant, while the ductility of the film could be tuned. The usage of comonomers that beared hydrogen bonding sites, allowed for the synthesis of polmyer-MMT films using butylacrylate as a monomer, which was up to this point impossible due to the extremely low glass transition temperature of PBA. Furthermore, it could be shown that this strategy resulted in composite particle films with an even higher pronounced reinforcement effect.

Chapter 7

Polymer-Nanosheet Coatings of PE-foils

Contents

Multilayer systems are omnipresent in high performance materials. For example, modern polymer-based food packagings are highly sophisticated layer systems consisting of an inner layer that is responsible for sealability, functional layers that introduce barrier properties for gasses, moisture or light, tie layers for adhesiveness, and an outer layer that provides mechanical stability and printability.[158] These highly complex composites are the result of profound research and are tailored for each individual application. One of the most fundamental component for inner and outer layer is PE that is one of the most commonly used technical polymers.[158] While PE provides excellent barrier properties against moisture, it does have poor barrier properties against oxygen, flavor, and aroma, all of which are necessary to keep the product fresh and consumer save.[158] Additional barrier layers enhance the materials performance in this regard.

Known ways to improve the barrier properties of foils are for example the usage of additional layers of proteins,[159,160] copolymers,[161] or nanoparticles.[158] MMT has been shown to provide an exceptional barrier against diffusion of gasses.[14,158,162] When arranged tightly in a "brick-and-mortar" structure, the only path for diffusion is around all particles resulting in a significantly increased mean path length that gas molecules have to travel to pass the additional layer.[158] One important topic when using nanoparticles for applications, is the possibility of migration of particles. Different studies have shown that well dispersed particles in a polymer matrix have a low to none probability to migrate. However, polymer coated nanoparticles should even show less migration out of polymer matrices due to their enhanced interaction with the surrounding polymer making them even saver for consumers.[163] Entanglements of the surface grafted chains with the matrix have to be disentangled for a particle to start moving through the matrix. The particles studied in this book should therefore be promising to produce consumer safe products.

In chapter 6 a biomimetic approach for the production of synthetic nacre-like self-supporting films was presented. *Ding et al.* have presented an approach for a nacre-like coating for poly(lactic acid) (PLA) and PVA foils using MMT.[14] They reported the successful formation of a tightly packed MMT coating on the PLA substrate that provided mechanical stability to the sample as well as enhanced barrier properties against oxygen and water and flame retardancy.[14]

In order to study the application of the composites that were developed in this book for coatings with enhanced mechanical properties and as barrier layers, low density poly(ethylene) (LDPE) foils were coated with polymer coated nanosheets. The different developed strategies from chapter 6 to modify the mechanical properties of matrix-free composites were adapted and PE-foils were coated with nanosheets grafted with either linear polymer (MMT-PMA$_{lin}$), a mixture of linear and star polymer (MMT-PMA$_{star}$), or polymer

Table 7.1.: Δ for PE foils coated with nanosheets that are modified with linear polymer, a mixture of linear and star polymer, or polymer containing hydrogen bonding sites (HB). Δ represents the mass of the entire coating made from polymer decorated nanosheets.

polymer modification	number of coating cycles	Δ / %
linear	1	0.77
star	1	0.79
HB	1	0.81
linear	3	2.62
star	3	2.83
HB	3	2.86

containing a small amount of hydrogen bonding comonomer (MMT-PMA$_{\text{HB}}$).

On an industrial scale, one of the simplest approaches to coat polymer foils is to dip coat the foils in a bath of the coating substance. On the laboratory scale, this can be replicated by simply dip coating the cut PE foils into a dispersion of the respective particles. The formation of a thin film of liquid on the surface of the substance after dip coating allows for self-assembly of the particles when the foil is hung vertically for drying.[14] The chosen solvent has to be a suitable solvent to disperse the particles and a good solvent for the grafted polymer. At the same time, it has to be a bad solvent for the PE foil. PGMEA meets all these requirements. Of each sample, coated foils with one coating cycles and tree coating cycles were prepared and analyzed. All polymer coated nanosheets that were used for these experiments had the same MMT content. Table 7.1 shows the percentage mass proportion (Δ) for each sample. It can be seen that coatings of comparable mass were obtained. As nanosheets with comparable MMT content were used for the coatings, it can be concluded that the MMT content of the coatings are comparable.

7.1. Mechanical properties

The main purpose of an MMT layer in these kind of systems are the barrier properties. However, it is always necessary to be aware of the influence this layer has on the overall mechanical properties of the composite because even tough traditionally rather the outer and inner layer are responsible for the mechanical properties, the MMT layer changes the properties of the entire composite and has to be accounted for when designing the material. To characterize the change in mechanical properties of the coated foils, tensile testing was performed. Representative stress–strain-curves are shown in Figure 7.1. The respective obtained values to characterize the mechanical properties are given in Table 7.2. Tensile testing shows clearly thermoplastic behaviour of all the samples indicated by a short linear elastic region followed by extensive plastic deformation until failure. It can also be seen that the coating does not change the overall performance of the material. The mechanical performance is therefore still mainly determined by the PE. However, when analyzing the stress–strain-curves in more detail (Table 7.2), it becomes clear that there is nonetheless an impact on the mechanical properties. The *Young's* modulus of the PE-foil increased for between 15% (coated once with MMT-PMA$_{\text{lin}}$ and 22% (coated three times with MMT-PMA$_{\text{HB}}$). This effect can be assigned to the MMT nanosheets that have already been shown to exhibit a *Young's* modulus one order of magnitude higher than the PE-foils (chapter 6).[41] The tensile strength seems to be slightly increased, while the strain at break remains constant within the margins of error. Especially the effect on the strain at break is clearly expected, as the plastic properties are determined by the PE-foil.

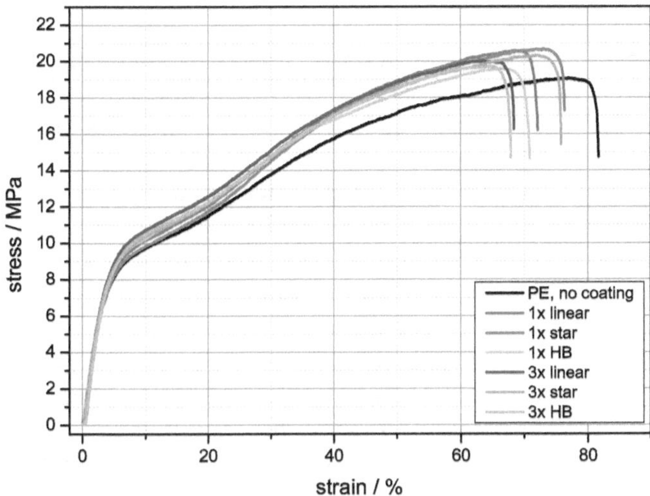

Figure 7.1.: Stress–strain-curves of used PE-foils and PE-foils coated with polymer coated MMT.

Table 7.2.: *Young's* modulus E, tensile strength σ_y, and stain at break ε_B of coated PE-foils.

sample	coating cycles	$E/$ N mm^{-2}	$\sigma_y/$MPa	$\varepsilon_B/\%$
uncoated	0	245 ± 8	19.5 ± 0.6	75 ± 6
linear	1	282 ± 7	20.3 ± 0.4	64 ± 7
linear	3	299 ± 17	20.7 ± 0.9	76 ± 7
star	1	296 ± 23	20.3 ± 1.1	71 ± 9
star	3	295 ± 6	20.5 ± 1.0	69 ± 7
HB	1	285 ± 37	20.5 ± 0.4	76 ± 2
HB	3	300 ± 14	20.1 ± 0.5	67 ± 8

7.2. Gas barrier properties

The barrier properties of the MMT-PMA coatings were analyzed using the self-designed setup shown in Figure 7.2. A sample is tightly screwed between two hollow metal blocks. The top part has an opening where gas from the environment can enter freely. The bottom part is attached to a manometer and a vacuum pump that can be tightly closed *via* a valve. When the foil is loaded into the setup, the pressure on the bottom side is reduced to 400 mbar. Then the valve between sample and pump is closed and the pressure is recorded over time.

From these experiments pressure–time-profiles can be recorded (Figure 7.3). The longer it takes for the pressure to reach 600 mbar, the lower is the gas permeability of the sample and the higher are the samples barrier properties. It is expected that the pressure in

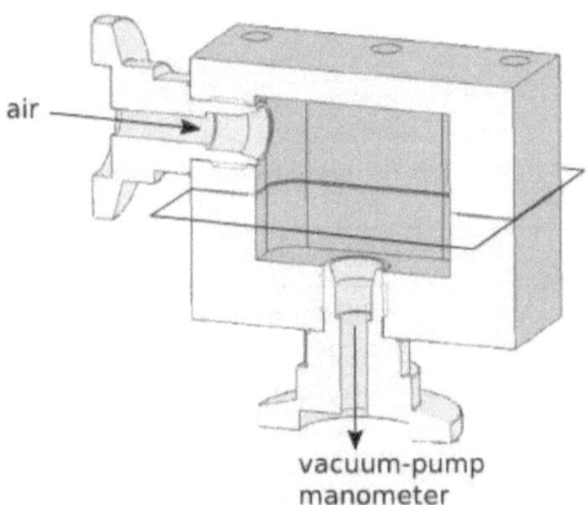

Figure 7.2.: Experimental setup for gas permeability measurements. Black outlines represent a PE-foil.

Table 7.3.: Averaged durations Δt for PE-foils coated with MMT nanosheets modified with PMA_{lin}, PMA_{star}, or PMA_{HB}. For visual clarity, only three error bars are shown. They are obtained as the standard deviation of three measurements.

sample	Δt
PE, uncoated	38 ± 11
1x linear	79 ± 11
1x star	74 ± 16
1x HB	77 ± 8
3x linear	86 ± 4
3x star	96 ± 8
3x HB	104 ± 8

the evacuated cell rises with time and will approach room pressure asymptotically. This behaviour is best observed for the samples that have the lowest permeability and do therefore take the longest time to reach 600 mbar, but can be observed for all the tested samples. Additionally, Table 7.3 shows the obtained average durations Δt for all tested samples. It can be seen that all coated foils show a significantly lower gas permeability then the uncoated PE-foil. Foils coated one time show a comparable gas permeability independent from the polymer-modification of the nanosheets. When coated three times, a trend of increasing barrier properties from coating with MMT-PMA over $MMT\text{-}PMA_{star}$ to $MMT\text{-}PMA_{HB}$ can be observed. A possible interpretation might be that through cross-linking in the polymer a denser and more compact coating is obtained. In a denser coating, the mean path length of molecules diffusing through it is increased, which can be observed as increased barrier properties or reduced gas permeability, respectively.

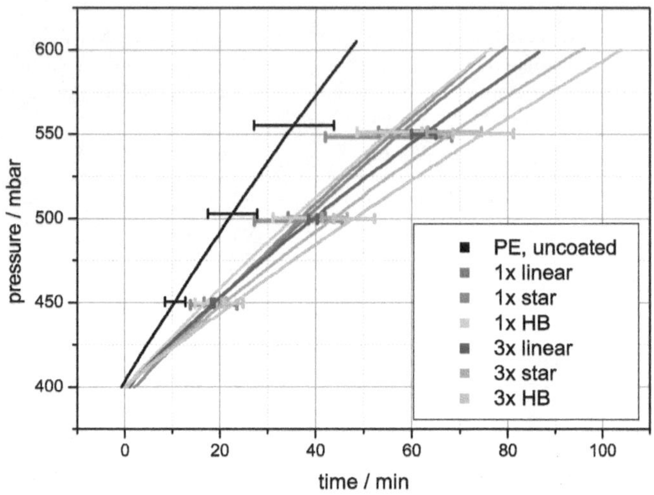

Figure 7.3.: Pressure–time-profiles of PE-foils coated with MMT-PMA$_{\text{lin}}$, MMT-PMA$_{\text{star}}$, and MMT-PMA$_{\text{HB}}$.

7.3. Conclusion

The successful coating of PE-foils with polymer grafted MMT has been presented. All samples show an increased *Young's* modulus compared to the uncoated PE foil. A setup to easily measure the gas permeability of coated foils was developed and used to analyze the barrier properties of different coatings. On one side of the foil the pressure was reduced and the time it took for the pressure to rise was recorded. It could be shown that a coating with polymer modified MMT nanosheets results in an increase of the barrier properties of the PE foils. Three different coatings were studied: nanosheets modified with either linear PMA, a mixture of linear and star PMA, or poly(MA-*co*-CEA), in which the CEA introduces hydrogen bonding sites to the polymer. Samples that were coated once with either of the nanosheets showed enhanced barrier properties. However, the gas permeability was independent from the type of coating. When the mass proportion of the coating was increased through repeated coating, it was observed that samples with cross-linked polymer modification of the nanosheets showed a lower gas permeability and consequently higher barrier properties. Foils coated with nanosheets modified with poly(MA-*co*-CEA) showed the lowest gas permeability.

Chapter 8

Concluding Remarks and Future Perspectives

Within this work, the synthesis of versatile polymer–layered-silicate-nanocomposites was presented. Based on the modification of organo-Montmorillonite using *grafting through* RAFT polymerization, three different types of nanocomposites were produced for diverse studies and applications:

- Conventional nanocomposites consisting of a polymer matrix with polymer modified MMT nanosheets. The impact of the nanofiller content and the grafted chain length on the mechanical properties of the composite material was studied in detail.

- "Matrix-free" composites of polymer coated MMT nanosheets. A reliable strategy for their production was developed. It was demonstrated that their mechanical properties can be tailored through variation of the grafted polymer's chain length and through cross-linking of the grafted polymer.

- PE-foils were coated using polymer decorated MMT nanosheets. It was showcased that this strategy significantly decreased their gas permeability.

In chapter 5 PMA coated MMT nanosheets were used to tune the mechanical properties of a PMA matrix. One strength of this approach is the precise tunability of the mechanical properties of the composite material i.e., the *Young's* modulus, tensile strength, ductility, and toughness by varying the MMT content and the chain length of the grafted polymer on the nanosheets. The studied effects were attributed to enhanced interactions between particles and matrix through entanglement of surface grafted chains with matrix chains and to the formation of a glassy layer around the particles that has a higher modulus and reduced chain mobility.

In order to further tune the composites properties, MMT nanosheets coated with block copolymers could be a very interesting approach. An additional inner block with a higher glass transition temperature has the potential to form an additional glassy layer which would provide an even stronger reinforcement for the material. The outer block – still consisting of the same polymer as the matrix – would still provide the miscibility of the surface grafted chains with the polymer matrix.

Additionally, in chapter 6 it was demonstrated that the introduction of star polymer or hydrogen bonding sites to the surface grafted polymer has an enhancing effect on the mechanical properties of "matrix free" composites. These strategies could be adapted for composites of a polymer matrix and nanosheet fillers. The grafting of star polymer would in this case be expected to result in enhanced entanglement of surface grafted and matrix chains. The grafting of polymer with the possibility to form hydrogen bonds would allow for a more stable filler network. It is literature known that conventionally high filler loadings are favorable for reinforcement of polymers because a network of fillers in the matrix is formed at sufficiently high filler content.[67,164,165] Using hydrogen bonding comonomers within the surface grafted polymer might be a valuable strategy to achieve the formation of a network at lower filler loadings as the grafted chains can form a more rigid network through hydrogen bonds than just through entanglements.

Polymer coated nanosheets were further used to prepare biomimetic, "matrix free", nacre-like composites that consist only of the silicate and the surface tethered polymer without any matrix. In chapter 6, it was shown that they exhibit exceptional mechanical properties which can be precisely tailored by variation of the surface bound polymer chain length and *via* introduction of chemical and physical cross-links in the form of star polymer and hydrogenbonding comonomers. In future experiments, another cross-linking strategy, the usage of reversibly cross-linkable monomers *via* UV irradiation, could be explored. This would possibly enable the production of films that can be moulded into a form while being in a non-cross-linked form, followed by irradiation with UV light in order to obtain a film that then has enhanced mechanical properties and will stay in the shape of the mould.

In chapter 7, it was shown that polymer grafted nanosheets can be used to coat PE-foils in order to increase their mechanical and gas barrier properties. It could be shown that with less than 1 % coating, the barrier permeability could be halved. When the mass percentage of the coating was increased to about 2.8 %, it could be demonstrated that nanosheets grafted with cross-linked polymer represent a more effective gas barrier than nanosheets grafted with linear polymer. Further studies on the gas permeability of specific gasses, e.g. oxygen or nitrogen, could be very intriguing for the application as coating in packaging materials for the food industry. By further optimizing the length of the grafted polymer and the filling content or tuning the topology of the grafted polymer, the performance of the gas barrier could be further improved. Furthermore, different polymers that inherently provide a better gas barrier than PMA can be used here.[41] So far, only dip-coating was applied in this book as a coating technique. Using other methods could further improve the macroscopic layer quality and the microscopic structure and arrangement of MMT. The detailed impact of the microscopic arrangement of the MMT nanosheets in this context could be studied using imaging techniques as for example SEM.

Beyond MMT, Halloysite would be an interesting candidate to investigate as nanofiller because of its one-dimensional nanostructure. Halloysite is an aluminosilicate clay like MMT but occurs as nanotubes with an outer diameter of 50 nm to 60 nm and a length of up to 10 µm. It is known that Halloysite can be modified using the same approach as used in this book, but can also be modified in order to enable *grafting from* RAFT polymerization.[166,167] Nanotubes are known to form composites with anisotropic mechanical properties[168,169] and would – especially in combination with MMT as a second filler – allow for the creation of a very flexible system for fabricating novel anisotropic composite materials.

Chapter 9

Experimental Section

9.1. Chemicals

DSPA, 4-Vinylbenzyl chloride (90%) and N,N-Dimethylhexadecylamine ($\geq$ 95%) were purchased from Sigma Aldrich and used without further purification. Azobis(isobutyronitril) (AIBN, $\geq$ 99%, Fluka) was recrystallized from methanol and stored at $-18\,°C$. Methyl acrylate (MA, Sigma Aldrich), Styrene (), Butylacrylate (BA, Sigma Aldrich) and Carboxyethylacrylate (CEA, Sigma Aldrich) were purified by passing through a column of basic alumina (Sigma Aldrich). APA was kindly provided by Dr. Jannik Wagner. Sodium Montmorillonite (MMT) was purchased from Alfa Aesar. Propylen glycol monomethyl ether acetate (p.a., J&K Scientific) and Dichloromethane ($\geq$ 95.5% (Sigma Aldrich) and d-2 (XX) were used as received.

9.2. Instrumentation

9.2.1. DMA

dynamic mechanical analysis (DMA) was performed on a PerkinElmer DMA8000. Temperature dependent measurements were performed in single cantilever mode with a strain of 0.3 mm, a heating rate of $2\,\mathrm{K\,min^{-1}}$ and a frequency of 1 Hz. Strain dependent measurements were performed in tensile mode. Material pockets were purchased from PerkinElmer.

9.2.2. SEC

size exclusion chromatography (SEC) was performed on an Agilent Technology 1260 Infinity System equipped with an isocratic high performance liquid chromatography pump, an auto sampler, a guard column (5x50 mm; 10 µm particle size, polymer standards service (PSS) styrene-divinylbenzene (SDV)) and three separation columns (PSS SDV, 10^6, 10^5, 10^3 Å pore sizes) and a detector system consisting of a UV-detector and a refractive index (RI)-detector. The system runs with THF as eluent at 35 °C and a flow rate of $1.0\,\mathrm{ml\,min^{-1}}$. The SEC was calibrated against polystyrene standards (M_n=3-3000 $\mathrm{kg\,mol^{-1}}$) of low dispersities (PSS GmbH, Mainz, Germany). All samples were dissolved in THF (c = $3\,\mathrm{g\,L^{-1}}$) containing Toluene as internal standard (c = $10^{-8}\,\mathrm{mol\,L^{-1}}$). Prior to injection, all samples were filtered with teflon filters (Th. Geyer GmbH, Renningen, Germany) of 0.45 µm pore size. All given SEC results are number average molecular weights obtained by evaluation of the RI signal.

9.2.3. NMR

^{1}H- and ^{13}C-NMR spectra were measured with a Varian Unity 300 at room temperature. As solvent d-2 Dichloromethane was used. The signal obtained from residual solvent protons was used as internal standard for all measurements. All spectra are base line corrected with MestreNova v10.0.2-15465 (Mestrelab Research SL, Santiago de Compostela, Spain).

9.2.4. EA

elemental analysis (EA) was performed by the analytical lab of the Institute of Inorganic Chemistry of Georg-August-University Göttingen on a 4.1 Vario EL 3 by Elementar.

9.2.5. TGA

Thermogravimetric analysis (TGA) was performed with a Netzsch (Selb, Germany) TG 209 F3 Tarsus over a temperature range from room temperature to $1\,000\,^\circ$C with a heating rate of $10\,\mathrm{K\,min^{-1}}$ under nitrogen atmosphere using Al_2O_3 crucibles, that were pyrolytically cleaned at $1\,000\,^\circ$C prior to each measurement.

9.2.6. SAXS

Small angle X-ray scattering (SAXS) measurements were performed in cooperation with Dr. Manuela Denz (Institute of X-Ray Physics, Faculty of Physics, Georg-August-Universität Göttingen) with the in-house setup XEUSS 2.0 (Xenocs, Sassenage, France). A Genix 3D source (Xenocs), Cu K_α radiation at a wavelength of 1.54 Å, 50 kV and 600 µA was used. The beam was focused to roughly 500 x 500 μm^2 using scatterless slits and multilayer optics. At a sample to detector distance of 575 mm, the scattered signal was recorded on a pixel detector (Pilatus3R 1M, 981 x 1043 pixels, pixel size 172 x 172 μm^2, Dectris Ltd., Baden, Switzerland). The primary beam intensity was blocked by a beamstop. All samples were measured for 2 h divided into 600 s intervals to check for radiation damage. All particles investigated were embedded onto a PMA support prior to the measurement. All composite samples were cast into roughly 1 mm thick testing samples. A PMA sample without MMT was measured as well and used for background subtraction to obtain the pure MMT signal. For data analysis, the 2D detector images were azimuthaly integrated using self-written Matlab scripts (Matlab2017a,

The MathWorks, Natick, MA, USA) based on the cSAXS Matlab base package, which is available at `https://www.psi.ch/en/sls/csaxs/software`. The integrated intensity $I(2\theta)$ was plotted against the scattering angle 2θ. SAXS data is shown in a range from $0.2°$ to $10.5°$, corresponding to real space length scales from $0.842\,nm$ to $44.1\,nm$. The recorded data were normalized to the exposure time and the transmission.

9.2.7. SEM

Scanning Electron Microscopy was performed by Volker Radisch (Institute of Material Physics, Faculty of Physics, Georg-August-Universität Göttingen) on a Nova Nanosem 650 in ultra high vacuum using an acceleration voltage of $3\,kV$ or $5\,kV$, respectively. All samples were coated with gold prior to the measurement to ensure conductivity of the sample.

9.2.8. ATR-FTIR

attenuated total reflectance Fourier-transform infrared (ATR-FTIR) measurements were performed on a Bruker Vertex 80 v. All samples were dispersed in PGMEA and drop-cast onto the measuring crystal. Samples were measured between $4500\,cm^{-1}$ and $450\,cm^{-1}$.

9.2.9. s-SNOM

Scattering scanning near-field optical microscopy (s-SNOM) measurements were conducted on a neaSNOM by neaspec GmbH (Munich, Germany) equipped with a near-field imaging system (optical detection modules for nano-imaging) including a near-field spectroscopy system (nano-FTIR). All samples were dispersed in PGMEA and drop-cast onto CaF-platelets. The solvent was evaporated at room temperature prior to measurement.

9.2.10. Tensile testing

Tensile testing was performed on a ZwickRoell Z2.5 tensile testing machine. The strain rate was chosen to be $20\,\mathrm{mm\,min^{-1}}$. The data were analyzed with TestExpertII (ZwickRoell, Ulm, Germany)and OriginPro 8.5G (OriginLab Corporation, Northampton, MA, USA). Average values given here are averages of at least three measurements. Stress stain curves presented show the dependency of the stress σ (σ=F/A), with F being the applied force and A being the cross section of the sample measured prior to testing, on the strain ε (ε=$(l - l_0)/l_0$), with l_0 and l being the length of the sample in the beginning and at any given time during the measurement, respectively. The Young's modulus E is obtained as the slope in the initial linear region. All measurements were performed at room temperature. Due to changes in ambient temperature, the temperature for different series may vary. It was ensured that all samples, that are compared to each other are measured at the same temperature.

9.2.11. TEM

Transmission Electron microscopy was performed using a Philips CM12 with an acceleration voltage of $120\,\mathrm{kV}$ and a resolution of $2.4\,\text{Å}$. Micrographs were taken using an Olympus CCD camera with a resolution of 1376x1032 pixels. The particles were dispersed in PGMEA ($1\,\mathrm{mg\,ml^{-1}}$) and dropcast onto TEM grids (Plano). The solvent was slowly evaporated at room temperature. All measurements were kindly performed by Dr. Wentao Peng.

9.2.12. Optical microscopy

Optical Microscopy was performed on a BX51-P Polarizing Microscope (Olympus, Tokio, Japan).

9.2.13. Gas permeability measurements

Gas Permeability Measurements were performed using a self designed setup (Figure 9.1). The film was placed in between the two halves of the device, which were then tightly sealed with screws. One valve was let open, while the other one was equipped with a manometer and a vacuum pump. This side was evacuated to about 400 mbar and the time that it took for the pressure to rise to a defined value (usually 600 mbar) was recorded.

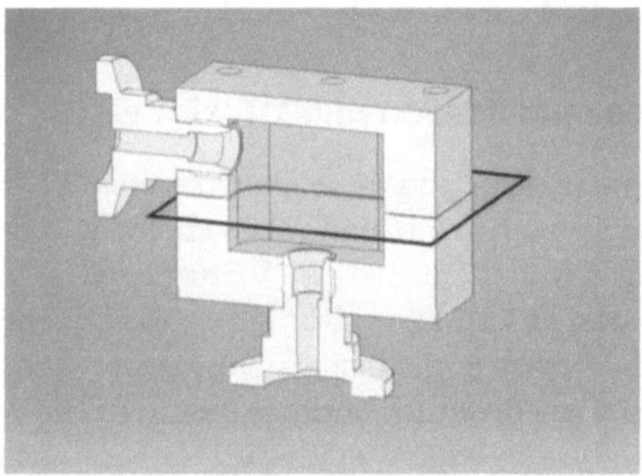

Figure 9.1.: Technical Drawing of the used setup to measure the Gas Permeability of coated polymer films. Black lines represent the position of the films.

9.3. Modification of MMT

9.3.1. Synthesis of VB16

4-Vinylbenzyl chloride (6.05 g, 0.040 mol, 1 Eq.) and N,N-Dimethylhexadecylamine (14.18 g, 0.053 mol, 1.3 Eq.) were dissolved in ethyl acetate (50 mL) in a round bottom flask. The solution was stirred over night at 40 °C. The solution was then cooled to −18 °C for 8 h, subsequently filtered and the precipitate was recrystallized from ethyl acetate. The product was dried *in vacuo* at room temperature and obtained as a white solid.

^{1}H-NMR (300 MHz) δ = 0.82-0.87 (m, 3 H, CH$_2$-CH$_3$), 1.20-1.29 (m, 26 H, CH$_3$-(CH$_2$)$_{13}$), 1.72-1.77 (m, 2 H, ((CH$_2$)$_{13}$-CH$_2$), 3.28 (s, 6 H, CH$_2$-N-((CH$_3$)$_2$)-CH$_2$), 3.46-3.50 (m, 2 H, (CH$_2$)$_1$3-CH$_2$)-N), 5.07 (s, 2 H, N-CH$_2$-Ph), 5.30-5.34 (d, 1 H, J=10, 8Hz, Ph-CH-CH$_{cis}$H$_{trans}$), 5.75-5.80 (d, 1 H, J=17.5Hz, Ph-CH-CH$_{cis}$H$_{trans}$), 6.62-6.72 (dd, 1 H, J=17.6 Hz, Ph- CH), 7.39-7.41 (m, 2 H, N-CH$_2$-C-(CH$_2$)$_2$-(CH$_2$)$_2$-CH), 7.58-7.61 (m, 2 H, N-CH$_2$-C-(CH$_2$)$_2$- (CH$_2$)$_2$-CH) ppm.

9.3.2. Surface modification of MMT with VB16

MMT (25.09 g) was dispersed in deionized water (1 L) and stirred for 3 days using a glass stirring bar and then cooled and the temperature was kept between 0 °C and 5 °C. VB16 (11.52 g, 0.03 mol) was dissolved in water (10 mL) and dropwise added to the dispersion of MMT. The temperature of the dispersion was kept at 0 °C to 5 °C and the dispersion was stirred for another 3 hours. The dispersion was then filtered and the residue was washed with deionized water until a test for chloride ions was negative. The particles were then dried *in vacuo* at room temperature followed by stirring in petrol ether (500 mL) for one hour. The particles were dried again to obtain VB16 modified MMT (MMT-VB16) as a grey powder.

9.4. Polymerizations

9.4.1. Polymerization of MA

In a typical polymerization procedure MA (2 g, 0.023 2 mol, 500 Eq.), DSPA (0.016 3 g, 4.66 · 10^{-5} mol, 1 Eq.), azobisisobutyronitrile (AIBN) (1.5 · 10^{-3} g, 9.13 · 10^{-6} mol, 0.2 Eq.) and toluene (50 wt.-%) were added into a polymerization vial. The solution was degassed with argon for 10 min at 0 °C. The polymerization was then conducted at 60 °C for 3 h and was ended by cooling the solution to 0 °C and exposure to air. The mixture was poured into aluminium dishes and the solvent and residual monomer was evaporated first at room temperature over night and then *in vacuo* at 100 °C.

All polymerizations of MA without particles were conducted based on this composition which is easily scalable. Molar masses of the polymers were controlled *via* variation of the reaction time.

9.4.2. Polymerization of MA in the presence of MMT-VB16

In a typical polymerization procedure, MMT-VB16 (0.1956 g) was weighed into a polymerization vial equipped with a stirrer bar. MA (2.850 g, 0.331 mol), AIBN (0.003 g, $1.66 \cdot 10^{-6}$ mol), DSPA (0.029 g, $8.28 \cdot 10^{-5}$ mol) and Toluene (2.850 g, 50 wt.-%) were added. The vials were degassed with Argon for 15 min and then polymerized at 60 °C for various times under constant stirring to yield PMA of various degree of polymerization. The polymerizations were stopped by exposure to air and dilution with cooled dichloromethane (34 ml). The polymer modified particles were separated from free polymer *via* centrifugation directly after polymerization. The free polymer was then poured into aluminum dishes and dried *in vacuo* in order to remove residual monomer and solvent and was then characterized *via* SEC. The particles were kept in dry state in order to prevent irreversible aggregation and were quickly used afterwards.

9.5. Synthesis of 3-arm star-RAFT agent

Triethylamin (3.8 mL, 27.26 mmol) was added to a solution of Propan-thiole (2.0606 g, 27.02 mmol) in Dichloromethane (250 mL) dropwise and was stirred for one hour. Carbondisulfide (12.94 mL, 214.5 mmol) was slowly added and the solution was stirred for another 15 min. Tribromide[152] (3.8759 g, 7.79 mmol) was added and the solution was

stirred for 72 h. Tribromide (Scheme 9.1) was kindly provided by
Dr. Wentao Peng. The product was washed with saturated $NaHCO_3$
(3×200 mL), hydrochloric acid (10%, 3×200 mL) and deionized water (3×100 mL). The organic phase was then dried over magnesium
sulfate and the solvent was evaporated under reduced pressure. The
product was purified *via* column cromatography with a mobile phase
of hexane/ethylacetate (10:1). The product was dried *in vacuo*.

^{1}**H NMR** ($CDCl_3$, 300 MHz): δ (ppm) = 5.32-5.20 (m, 1H), 4.89-4.75
(m, 3H), 4.46-4.13 (m,4H), 3.34 (t, ^3J=7.2 Hz, 6H), 1.75 (h, ^3J=7.4 Hz,
6H), 1.59 (d, ^3J=7.0 Hz, 9H), 1.02 (t, ^3J=7.4 Hz, 9H).

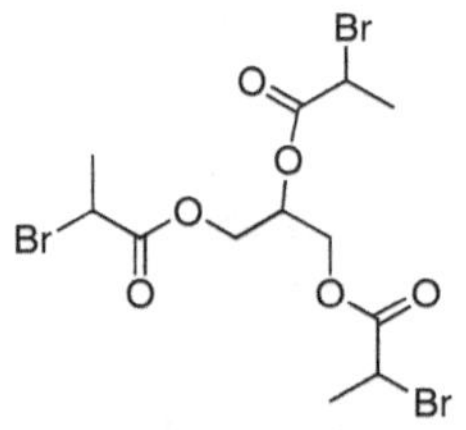

Scheme 9.1: Structural formula of tribromide. Precursor for the synthesis of
S3. Tribromide was kindly provided by Dr. Wentao Peng.

9.6. Polymerizations for the production of MMT-PMA and MMT-PBA films

All polymerizations were carried out as described in subsection 9.4.2.

9.6.1. Polymerizations for the production of MMT-PMA films with varying chain length of grafted polymer

Table 9.1, Table 9.2 and Table 9.3 give the chosen compositions and
reaction conditions for the synthesis of films with PMA of varying
chain length. For samples **L-I** to **L-IV** the same composition of the

Table 9.1.: Composition of polymerization solutions for the synthesis of films **L-I** to **L-IV**. A stock solution of AIBN (0.020 g) in toluene (2 mL) was prepared and 0.27 ml was added to each vial. The polymerizations were conducted at 60 °C. Polymerization durations are given in Table 9.2.

chemical	m / g	n / mol	eq.
MMT-VB16	0.196	-	-
MA	2.850	0,0331	400
DSPA	0.029	$8.27 \cdot 10^{-5}$	1
toluene	2.850	0,0635	-

Table 9.2.: SEC results of the polymerizations for the synthesis of films **L-I** to **L-IV** and respective polymerization durations.

sample	grafted $\overline{M_n}$ / $10^4 \mathrm{g\,mol^{-1}}$	duration / h
L-I	0.4	1
L-II	1.2	2
L-III	1.8	2:20
L-IV	2.2	3

solutions was chosen while polymerization time was varied in order to achieve different conversions. For samples **L-V** to **L-VIII** the monomer to RAFT ratio was varied as indicated.

9.6.2. Polymerizations for samples prepared *via* mixing approach

Polymerizations were conducted analogous to compositions given in Table 9.1. For nanosheets grafted with short chains the duration was chosen to be 1 h. For nanosheets grafted with short chains the duration was chosen to be 3 h.

For film preparation the particles were dried after polymerization

Table 9.3.: Compositions and respective SEC results of the polymerizations
for the synthesis of films **L-V** to **L-VIII**. Polymerizations were
conducted for 3.5 h at 60 °C.

	L-V	**L-VI**	**L-VII**	**L-VIII**
m(MMT-VB16) / g	0.196	0.196	0.195	0.196
m(MA) / g	1.210	3.563	4.275	5.002
m(DSPA) / g	0.029	0.029	0.029	0.029
m(toluene) / g	1.809	3.566	4.284	4.997
$\overline{M_n}$ / 10^4 g mol^{-1}	2.8	3.2	3.9	4.3

and combined in the chosen ratios. They were thoroughly redispered
in PGMEA prior to film production as described in chapter 6.

9.6.3. Polymerizations for samples prepared with a mixture of linear and star RAFT agent

Table 9.4 gives the fractions and molar amounts of the two different
RAFT agents used for the preparation of films with mixtures of linear
and star RAFT. Table 9.5 gives the composition for the other components, which were identical for each polymerization. A stock solution
of AIBN (0.065 g) in toluene (6.6 mL) was prepared. Table 9.6 gives
the exact amounts of stock solution added to each polymerization.
Amounts of AIBN were calculated based on a constant proportion of
1:0.2 of RAFT groups to AIBN.

9.6.4. Statistical copolymerization of MA with CEA in the presence of MMT-VB16

Copolymerization of MA with CEA in the presence of MMT-VB16
were conducted by weighing 0.196 g MMT-VB16 into a polymerization vial that was equipped with a stirring bar. Stock solutions of MA,
toluene, CEA, AIBN, and DSPA (Table 9.7, Table 9.8) were prepared

Table 9.4.: Fractions and molar amounts of linear (DSPA) and star (S3) RAFT agent.

sample	x_{DSPA} / %	x_{S3} / %	n_{DSPA} / 10^{-5} mol	n_{S3} / 10^{-5} mol
S-I	70	30	5.80	2.49
S-II	60	40	4.91	3.28
S-III	50	50	4.14	4.16
S-IV	40	60	3.34	4.98
S-V	20	80	1.66	6.64
S-VI	0	100	-	8.24

Table 9.5.: Composition of polymerizations with varying amount of S3 and DSPA.

chemical	m / g	n / g
MMT-VB16	0.196	-
MA	2.850	0.0331
toluene	4.275	0.0464

and 5.515 g of the solution was added to each vial. The polymerization vials were degassed with argon for 15 min. Polymerization times

Table 9.6.: Volume of AIBN stock solution added to each polymerization vial for polymerization with varying amount of S3 and DSPA.

sample	$V_{stock\ solution}$ / mL
S-I	0.43
S-II	0.49
S-III	0.54
S-IV	0.60
S-V	0.71
S-VI	0.83

Table 9.7.: Stock solution composition for the copolymerization of MA with CEA in the presence of MMT-VB16 resulting in samples with 7 wt.-% CEA content. The polymerization time was 3 h at 60 °C.

chemical	m / g	n / mol	eq.
MA	11.309	0.131	380
toluene	11.309	0.123	-
CEA	0.947	$6.57 \cdot 10^{-3}$	20
AIBN	0.011	$6.82 \cdot 10^{-5}$	0.2
DSPA	0.121	$3.44 \cdot 10^{-4}$	1

Table 9.8.: Stock solution composition for the copolymerization of MA with CEA in the presence of MMT-VB16 resulting in samples with 3 wt.-% CEA content. The polymerization time was 3 h at 60 °C.

chemical	m / g	n / mol	eq.
MA	9.217	0.107	392
toluene	9.217	0.100	-
CEA	0.315	$2.18 \cdot 10^{-3}$	8
AIBN	0.009	$5.48 \cdot 10^{-5}$	0.2
DSPA	0.096	$2.72 \cdot 10^{-4}$	1

are given in the respective tables. The polymerizations were ended by exposure to air and dilution with cooled DCM. The free polymer was separated from the particles *via* centrifugation and dried *in vacuo* over night at 100 °C. Conversion was measured gravimetrically by determination of the mass of the free polymer and the increase in mass of the particles.

Table 9.9.: Composition of statistical copolymerizations of BA and APA in the presence of MMT-VB16.

chemical	HB-III	HB-IV	HB-V	HB-VI
m(MMT-VB16) / g	0.196	0.196	0.196	0.196
m(BA) / g	2.817	2.749	2.682	2.615
n(BA) / mol	0.0219	0.0215	0.0209	0.0201
m(APA) / g	0.058	0.172	0.287	0.401
n(APA)/ 10^{-3} mol	0.262	0.784	1.31	1.83
m(DSPA) / g	0.0194	0.0194	0.0194	0.0194
n(DSPA)/ 10^{-5} mol	5.56	5.56	5.56	5.56
m(AIBN) / 10^{-3} g	1.82	1.82	1.82	1.82
n(AIBN)/ 10^{-5} mol	1.11	1.11	1.11	1.11
m(toluene) / g	2.85	2.85	2.85	2.85

9.6.5. Statistical copolymerization of BA with APA in the presence of MMT-VB16

BA, APA, AIBN, DSPA, and toluene were weighed into polymerization vials (Table 9.9) and degassed for 15 min with argon. The polymerization was conducted at 60 °C for 6 h.

www.ingramcontent.com/pod-product-compliance
Lightning Source LLC
LaVergne TN
LVHW042202190726
843493LV00006B/1781